Effective Audio-Visual
A User's Handbook

Effective Audio-Visual

A User's Handbook

Robert S. Simpson

Focal Press
London and Boston

Focal Press
is an imprint of the Butterworth Group
which has principal offices in
London, Boston, Durban, Singapore, Sydney, Toronto
and Wellington

First published 1987

© Robert S. Simpson, 1987

British Library of Cataloguing in Publication Data

Simpson, Robert S.
    Effective audio-visual: a user's handbook.
    1. Audio-visual materials
    I. Title
    001.55'3   LB1043

ISBN 0-240-51254-5

Library of Congress Cataloging in Publication Data

Simpson, Robert S.
    Effective audio-visual.

    Includes index.
    1. Audio-visual equipment.   I. Title.
TS2301.A7S57   621.38'028'2   87-12107
ISBN 0-240-51254-5

Photoset by Butterworths Litho Preparation Department
Printed and Bound in Great Britain at the University Press, Cambridge

Preface

Audio-visual has come to mean many different things to different people. To some the use of a single slide is audio-visual, to others only a program using the latest video technology qualifies for the epithet. Technology itself has made the whole field more intimidating, and some potential users could be forgiven for either giving up the idea of using audio-visual at all because it seems so complicated, or worrying whether they have chosen the right medium.

This book is written for the audio-visual user. This is a wide public, because audio-visual media can be used by any organization. Usually the person who has to use or commission AV methods has no formal training in their use; his or her job is often some functional role within the organization for which AV is seen as useful support.

The aim of the book is to help these people in two ways. Firstly to convey *advice* on the choice of AV medium and how to prepare programs and materials, and secondly to convey *facts*, especially about equipment, presentation methods and environments that will help in the practical implementation.

It might be argued that someone who has to commission AV only needs advice on choice of medium, and that technical matters can safely be left to technicians. Unfortunately, many good audio-visual shows have not realized their full potential because the sponsor has failed to take any account of the environmental needs of AV. A technician cannot reverse a management decision to give presentations in a totally unsuitable room. On the other hand the manager with responsibility for the success of a presentation who also understands the basic principles involved can ensure that the problem never arises in the first place.

Because the book includes factual information it will also be of use to those already in the audio-visual business, whether as program producers, staging technicians or professionals in one branch of AV who need basic information about another.

The chapters are, as far as possible, complete in themselves. Each is a review of a particular subject, so can be read on its own.

Audio-visual methods have a valuable contribution to make in education, training, government, art, culture, commerce and industry. Their use can be stimulating, effective and fun. This book is intended to help users make the most of them.

R.S.S.

Acknowledgements

Parts of this book have previously been published in a house publication of Electrosonic Ltd entitled *What is Audio-Visual?*.

In the text there are necessarily references to company trade marks. When known these have been indicated as trade marks: for example, Carousel^{TM}, a trade mark of Kodak Ltd and Eastman Kodak Co.

Illustrations of audio-visual equipment are representative. The inclusion or exclusion of particular products does not imply any special endorsement or adverse comment. The author acknowledges help with illustrations from many colleagues in the audio-visual business, and the source of illustrations is indicated in the captions. Unattributed photographs and drawings are by Electrosonic Ltd, David Nuttall and the author.

Contents

List of tables

Chapter 1

The choice of audio-visual media

The first confusion that must be faced is the distinction between 'Audio-Visual' and 'Visual Aids'. This book covers both, because they are generally used together, and often use the same equipment. For the purposes of definition audio-visual is taken to mean a medium using recorded sound, whereas visual aids accompany a live presenter.

A summary of visual aids is shown in Table 1.1.

Table 1.1 Visual aids

Display material, including product display

Flipchart

Overhead projector

Writing board

Magnetic and other display boards

Slides

Filmstrip

Video display of single images

Video display of computer output (graphics and text)

It is clear that some of the aids require more preparation than others, and that some permit writing on by the presenter. A similar summary for audio-visual is shown in Table 1.2.

The problem is deciding which medium to use. This chapter reviews the priorities related to business and training presentations. The special considerations required for exhibitions, big audiences, and permanent installations are more fully dealt with under separate headings.

When choosing a presentation medium it is important to concentrate on the particular presentation objectives. The effectiveness of the presentation is more important than the means

Table 1.2 Audio-visual

Super 8 movie

16 mm movie

35 mm movie

70 mm movie

Filmstrip with sound

Single slide with sound

Dissolve slide with sound

Multi-image slides with sound

Videotape/videodisk displayed by monitors

Videotape/videodisk displayed by projection

Multi-screen video

Computer output display linked to sound

Sound and light technique

Mixed or multi-media

employed. The specialist needs of a limited audience are very different from those of a publisher of video programs or a producer of public entertainment films, which means that there is often an argument for using the simplest possible medium. However, the discipline that a medium imposes can also greatly improve the effectiveness of a presentation.

Often a user finds that more than one medium is needed – usually a simple visual aids system and a separate recorded program system. The choice of medium for a particular application depends on the answer to a series of questions. Unfortunately, there is not always a clear-cut answer to each, but there is a 'favorite' answer. The actual application will then point to one medium as the best single answer to the problem. This is illustrated in Table 1.3.

Table 1.3 Choice of visual aids for groups

What is best if . . .	Favorite answer
'Once only' material is needed for an informal audience?	Flipchart/overhead projector (OHP)
A 'teaching' session is needed?	Writing board/OHP
A mixture of frequently used and specially prepared material is needed?	OHP
The visual must be changed while being displayed?	OHP/magnetic board
The visual must match AV material in impact and quality?	Slides
It must be easy to prepare material 'in house'?	OHP
Daylight is present?	Writing board/ OHP/flipchart
On-line computer data must be displayed?	Video or data projector
A prestige presentation is to be given?	Slides

Table 1.3 reviews the choice of visual aids for groups, but does not answer all questions. The occasions when visual aids are used vary greatly in formality and importance. A group of engineers having a discussion in a laboratory can manage very well with a writing board as their main visual aid; a monthly board meeting of a major corporation may in some ways resemble a discussion meeting, but the speed at which discussions can be conducted will depend greatly on how well information is presented to the group.

At this point a 'law' applies; the effectiveness of the visual aid can be proportional to the time taken to prepare it. A well-prepared set of slides presenting financial data to a board meeting may take longer to make than a set of figures hastily duplicated and handed round. However, the presentation of financial data to a group by slides is far more satisfactory than the use of handouts because:

☐ Everyone is looking at the same data and is not reading ahead.

☐ The discipline of preparing the data in a form that can be read by a group ensures that it is easier to assimilate.

☐ The formality imposed by a standard medium speeds up the presentation and discussion process.

Chapter 2, on the preparation of visuals, demonstrates that the production of good visuals is not necessarily more time-consuming than the production of bad ones. In recent years many difficulties have been removed. Lettering machines that are quick and easy to operate mean that it is no longer necessary to choose between typewriting, which is difficult to read and the delay and expense of typesetting. The introduction of the high-quality instant slide allows a uniform presentation quality, even when material must be prepared at the last minute.

When it comes to audio-visual it is necessary to make a distinction between individual and group communication, because the equipment used for the different applications gives a different emphasis to both the questions and the answers. Table 1.4 and Table 1.5 review the two fields.

The tables raise a few questions. For example, video is suggested less often than might be expected. This is because most serious business presentations depend heavily on good-quality *still* images (such as charts, graphics and product photographs) that must be presented on a large scale. The normal TV screen can show only a limited amount of information when viewed by an audience. If video projection is used the results are

Figure 1.1 The writing board goes electronic. Anything written on the Muirhead Copyboard can be reduced to A4 size and distributed as a paper copy.

Table 1.4 Choice of AV medium for individual communication

What is best if . . .	Favorite answer
A lot of copies are needed?	Video/filmstrip
Only a few copies are needed?	Slides
Easy program making and editing are required?	Slides
Movement is required?	Video/Super 8
Good image quality is required?	Slides
Portability is required?	Video/filmstrip/slides
Low production costs are essential?	Slides
Interactive programs are required?	Videodisk

Table 1.5 Choice of AV medium for group communication

What is best if . . .	Favorite answer
Wide distribution of the program is required?	16 mm/video
Good image quality is required?	Slides
Good sound quality is required?	Slides/video
Easy program editing is required?	Slides
Movement is required?	16 mm/video
A large audience is required to be motivated?	Multi-image
A live presenter also takes part?	Slides
Production costs must be reasonable?	Slides

Starting in business can be easy . . .

— or a trifle difficult . .

. . . But whatever your business problems . .

. . . The Bank are always there to Help.

Figure 1.2 The storyboard is the key to any audio-visual program. One image, one idea.

acceptable for moving images in the dark, but are not really good enough for large still images with some room lighting.

There is no doubt, however, that *small* group presentation is making increasing use of video, especially in conjunction with computer-generated graphics and text displays. This, in turn, requires some new thinking about how information should be presented, because the amount of information that can be displayed on one image is significantly reduced.

Figure 1.3 The East Asiatic Company's presentation room/boardroom in Copenhagen benefits from good lighting which allows proper presentation of visual aids and AV programs. (Photo Lys and Lyd A/S.)

Figure 1.4 An unusual application of multivision. A 48-projector system backs a scientific general-knowledge quiz called Head to Head on German television. (Photo Alexander von Cube/WDR.)

As users of audio-visual methods gain experience, they may well find it necessary to use a variety of media and presentation methods. Thus, if a lot of use is made of video sequences with moving images for training, the use of slides or overhead transparencies for the still images (as opposed, for example, to computer-generated images presented on the video system) can be a positive advantage because this helps vary the pace of the presentation.

All AV methods require some preparation, not only in their initial production, but also in their actual use. Obviously a special occasion will justify special effort and perhaps extra expenditure. But it is in day-to-day use that AV can be most effective, and it is therefore important to plan from the outset how material is going to be prepared. It has become increasingly possible to use one medium for production and initial showing, and another medium for distribution. The next three chapters develop this theme.

Plate 1 Above: examples of computer-generated slides that show the versatility of slide-making systems. (Slides by The Image Bureau, London)

Plate 2 The London Experience is an example of multimedia audio-visual being used for tourist and visitor shows.

Plate 3 The History Pavilion at Expo 85, Japan, featured 'Rice and Iron', a multi-image show of great size and impact. It used more than 80 xenon arc slide projectors.

Plate 4 The German Pavilion at Expo 85 used multi-image to create a lively environment.

Plate 5 Lasers are often used to give spectacular results at new product launches. (Photo Laserpoint Ltd)

Plate 6 The Evoluon at Eindhoven has pioneered interactive exhibits. (Photo James Gardner Studio)

Plate 7 Above: modern planetaria combine the use of traditional star projection with multiple slide, effects and lighting control. At the McLaughlin Planetarium in Toronto, 120 slide projectors, 640 effects circuits and 75 dimmers are under computer control.

Plate 8 Reuters need to explain their specialist communication services to small client groups. Their presentation room is equipped with dual video/data projection, random access slide projection and comprehensive lighting and audio control.

Plate 9 The videowall principle can be applied to displays of any size. Here a giant 144-monitor wall is used in a TV quiz show. (Photo courtesy ZDF Munich and BKE Bildtechnik)

Plate 10 Examples of videowalls in public display: a 35-monitor wall sponsored by United Technologies and installed at EPCOT, Florida (top) and a 27-video projector wall installed in the NEC pavilion at Expo 85 (above).

Plate 11 London Transport Advertising were one of the first users of videowalls for advertising. This display is at their London Airport Heathrow underground station.

Chapter 2

The preparation of visuals

Some years ago Kodak Ltd in the UK put out a leaflet entitled *Let's stamp out awful lecture slides*. It began:

> We are organizing a crusade. Nothing controversial, of course. We are not party politically minded, and we don't feel too strongly about the Corn Laws. However, what we do get rather steamed up about are *awful lecture slides*.
>
> You know what we mean. An awful slide is one which contains approximately a million numbers (and we've left our opera glasses behind). An awful lecture slide is one which shows a complete set of engineering drawings and specifications for a super-tanker. An awful lecture slide is one which shows about two dozen dials when only one is necessary.
>
> That's what we're crusading against, and we would like you to join us. Of course we know *you* don't produce that sort of slide, but you probably know somebody who does. Do him a favor and pass on these tips . . .

For 'lecture slides' the leaflet could just as well have read 'overhead transparencies' or even 'flipcharts'; the problem is the same for any text-based visual presented to an audience. So before deciding on which medium to use it is important to understand the ground rules that apply to the preparation of *all* kinds of visual.

Legibility

The biggest single error made in preparing visuals is the mistaken idea that legibility in one form means legibility in another. A printed page is read at a distance of 30–50 cm (12–20 in). In a lecture theater a slide or overhead transparency that is projected onto a 2 m (6 ft) wide screen may have to be viewed from a distance of 20 m (60 ft). Reading text in this way is like reading this book from a distance of 3 m (10 ft).

Thus in AV terms it is just as much a crime to make an overhead transparency directly from a typewritten page, as it is to photograph that page and make a slide from it. To produce a legible visual requires preparation, and in principle there is little difference in the time needed to produce a usable slide, overhead transparency, flipchart page or computer text page. The decisions as to which to use depends, therefore, on other factors.

Figures 2.1, 2.2 and 2.3 demonstrate how lettering should be sized for the different methods of presentation. It is no coincidence that they all end up by limiting any one visual to about the same amount of text.

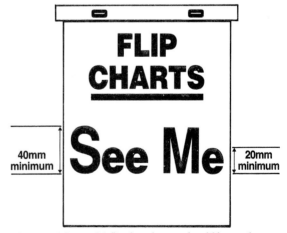

Figure 2.1 On an A2 flipchart letters should be not less than 20 mm (¾ in) high for a maximum viewing distance of 8 m (26 ft). Longer viewing distances mean a bigger chart and larger letters.

Clutter

The booklet *Slide Rules*, by Antony Jay, that accompanies the splendid Video Arts – Kodak Ltd co-production *Can we please have that the right way round?* suggests that a good rule is never to put more text on a slide than would be printed on a T-shirt. This may be taking matters a little further than is practical, but does make the point that any

visual should be kept simple. It is not just a matter of limiting the amount of text, but of considering *any* image in terms of how it stands up to presentation.

Thus, an engineering drawing or circuit diagram may be ideal for printing in a book for individual study, but be quite useless as a slide or overhead transparency. This is not just because the audience will have difficulty in deciphering the details;

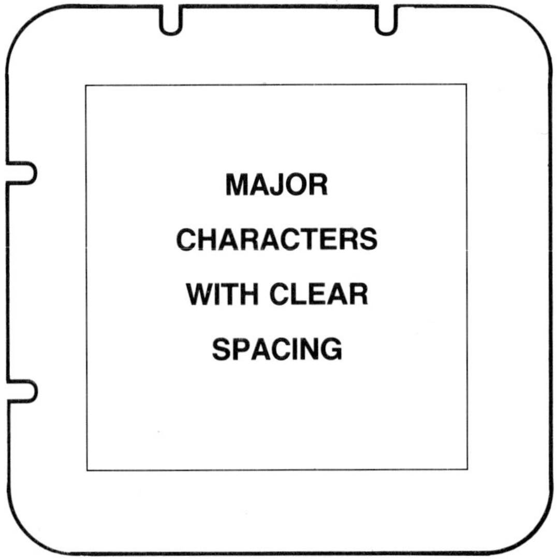

Figure 2.3 The rules for preparing text slides are the same as for overhead transparencies.

Figure 2.2 On an overhead transparency there should be a maximum of fifteen lines. Major characters should be one space high, and there should be a clear space between each line.

Figure 2.4 . . . therefore it is not surprising that the same rule applies for text on a video screen.

while they are busy screwing up their eyes to examine the details of the visual, they are not listening to the presenter.

So the next rule is never to put more on a visual than is necessary to make the point being discussed. In fact, it is possible to go further and suggest that the best visuals are those that are not quite complete and rely on the presenter to complete the picture.

The specific advice which emerges from these general observations applies to all kinds of prepared visuals, whether slides, overhead transparencies, flipchart pages, or computer pages intended for video projection. Some of the points demonstrated in the accompanying illustrations, in no particular order, include:

☐ Keep the number of words to a minimum.

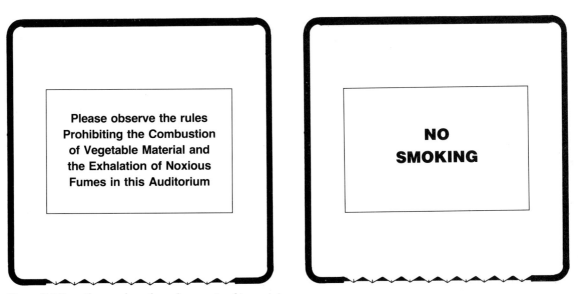

Figure 2.5 Keep the number of words on a visual to a minimum.

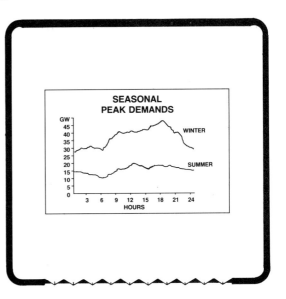

Figure 2.6 Avoid tables of figures. Use a chart or graph.

☐ Never show large amounts of text which allow the audience to read ahead of the presenter.

☐ Remember that a visual is a support, not a substitute, for the presenter.

☐ Always simplify diagrams to the essentials.

☐ Avoid big tables of figures. Either break them down into smaller tables, or, better still, convert them to a graph, bar chart or pie chart.

☐ When labelling diagrams or charts use only horizontal lettering – *never* angled lettering.

☐ Use color wherever possible to clearly distinguish between sections of a diagram or sectors of a chart.

☐ When using color, be careful to use combinations that complement each other. Some background/foreground combinations, e.g. pink letters on a red background, can be illegible when projected.

☐ When photographs are used, show only the essential. This may mean reproducing only part of an existing picture to eliminate irrelevant background.

☐ Make sure that all visuals for a presentation are the same format. Do *not* mix 'portrait' and 'landscape' formats.

☐ Never put an image in front of an audience that is not directly relevant to what is being said by the presenter.

☐ Do not bring on an image too early, and, most importantly, get it off the screen as soon as it has been discussed.

☐ Wherever a complex image is required, try to break it down into several simple images. If this is not possible, show it as a build-up sequence.

The last point is the most important of all. The technique of 'successive reveal' or 'build-up' helps communication because it ensures that the audience are only presented with one new piece of information at a time. The presenter is forced to keep to a logical order of presentation, and neither he nor the audience can jump ahead, because they do not see visual information on the screen before it is required. Although the principles of visual design are the same for all the standard media, it is the way in which image *sequences* can be presented that affects the choice of medium.

People who give or organize presentations on a regular basis need to develop methods of making or commissioning visuals. It is important to understand that even if users do not intend to make the visuals themselves, they must be aware of the principles of good visual design. A

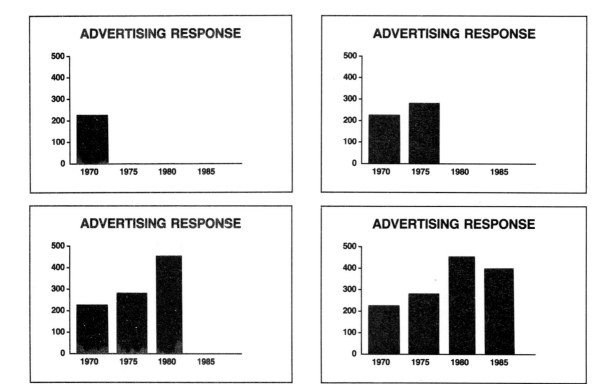

Figure 2.7 Use a build-up sequence if the final chart is complex.

slide-making bureau will produce the slide asked for – if the slide has too much text, it is the fault of the commissioner, not the bureau.

Making visuals

Now it is time to review different ways in which visuals can be made. At this stage the methods will be described without relation to their cost or suitability for in-house use. In many cases there are equally effective alternatives, and it can simply be a matter of personal preference as to which is used. For example, people with an aptitude for using personal computers are likely to want to use them to make visuals; others, who hate the sight of computers, will probably be much happier with simple lettering machines.

Common sense must be applied to some of the methods listed. For example, it may seem strange to suggest that one of the ways in which overhead transparencies can be made is directly from a slide. Why not use the slide itself and project that? If a presentation consists mainly of photographic images the best way to present them *is* by slides; but if it is given in a daylit room and most of the material is conventional overhead transparencies, it can make sense to convert the one or two photographic images required into big overhead transparencies.

Overhead transparencies

The 'fuel' for the overhead projector (OHP) is the overhead transparency. This is a piece of transparent film, with or without a card border, usually 25 cm × 25 cm (10 in × 10 in) or A4 size. Although the transparency is sufficiently big that it is possible to prepare the visual directly on the transparent material, a copying process is usually included as part of the preparation.

It is possible to write directly on the transparent film using a felt-tip marker pen. Usually this is done only when the presenter is using the OHP as a writing board, but sometimes it is appropriate to prepare a sequence of hand-written transparencies in advance.

Better-looking transparencies use proper lettering and even artwork. Although it is possible to apply this directly to the transparent film, it is more usual to prepare the initial material as artwork on plain paper, and then make a transparency from it.

The transparency is made in one of two ways. Best results are obtained using thermal or infra-red films. The original is placed against the film and

the two are passed through a thermal copier. The resulting transparency can then be:

☐ black on clear;
☐ black on color;
☐ color on clear;
☐ clear on color;

depending on the type of film used. The range of colors is typically limited to blue, red, green and yellow.

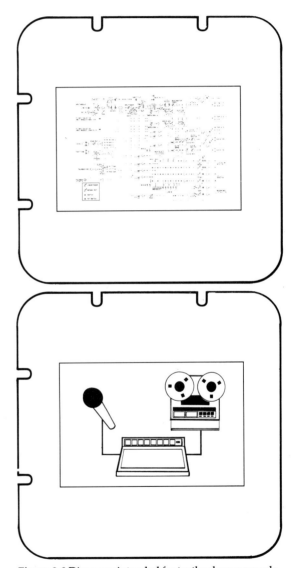

Figure 2.8 Diagrams intended for textbooks or manuals are too complicated for group presentation. They should be simplified.

Figure 2.9 Good-quality lettering for the in-house production of visuals can be produced with a lettering machine. (Photo Kroy Europe Ltd)

An alternative method, which is limited to black on clear or black on color tint, is to use a standard plain-paper copier. This can either be the dry toner type (e.g. Xerox™) or the liquid toner type (e.g. Nashua™). Here the plain-paper copier (PPC) film is placed in the copier's feed tray and the original placed on the machine exactly as if an ordinary copy were being made. An ordinary copy should be made first – if the original makes a good paper copy, then it should also make a good transparency.

Lettering on the original can be prepared in several different ways. Typewriting is often used, but it should definitely be avoided unless a display typewriter is available. Ordinary typewritten characters do *not* meet the requirements for legibility. Rub-on lettering such as Letraset™

Figure 2.10 Text and diagrams prepared to the appropriate size on a personal computer can be used directly to make an overhead transparency. Here the printer of an Apple Macintosh is printing out simple graphics created on the computer. (Photo Apple Computer UK Ltd)

gives good results, especially when additional standard symbols are needed. Much quicker to use are lettering machines that prepare the lettering in aligned strips.

The advent of the personal computer has introduced another highly satisfactory way of preparing OHP transparencies. However, it is not good enough to use just the standard letter printing because this would be no better than typewriting. What is needed is a computer with a graphics printer, and with a program that allows the creation of page layouts using large fonts and, possibly, diagrams and charts. In this respect a computer like the Apple Macintosh™ is outstanding.

Color can be added to overhead transparencies in several ways. Apart from the options offered by the basic film, more elaborate coloring can be achieved by:

☐ hand tinting with marker pen;

☐ hand coloring by cutting and pasting on color adhesive films;

☐ mounting several separate color-on-clear originals together in the same mount.

If more elaborate coloring is required it is best to use standard photographic methods. Full-color transparencies for OHP use can be made from any color original, whether flat artwork, a colour transparency or a photograph in another size. For those in a hurry, Polaroid make an instant OHP transparency film.

Slides

Although there is no reason why an OHP transparency should not look professional, they are normally used for less formal occasions. However, if complete flexibility in image make-up, a full range of colors and the highest possible image quality are required, then slides are called for.

Lettering for slides is prepared in the same way as for overhead transparencies. Typing can be used because it is possible to blow the lettering up to a legible size. However, because this then reveals a poor letter quality, typing tends to be used only for effect. The user requiring only a few slides will use rub-on lettering or a lettering machine. The professional slide maker will use properly typeset lettering, either made on his own machine or by a local printer. It is only necessary to have the letters set with the correct relative size, and in high-contrast black and white, since coloring and positioning can all be done by the camera.

The rostrum camera

The enormous flexibility of slides, both as visual aids and as components of AV shows, is due to the way that photographic images can be manipulated. For AV work the secret behind this manipulation is the rostrum camera. Many users do not need the full facilities of a professional rostrum camera; their needs can be adequately met by a simple copy stand. Others do not need all the features of the large machines. However, a brief description of a typical camera serves to illustrate both what it is and what it can do.

The main component of a rostrum camera is a rigid table. Flat artwork, which can be any size from about the size of a slide up to 1.2 m (4 ft) wide, can be placed on the table. The artwork can be top-lit by suitably placed lighting, which usually has polarizing filters to eliminate glare. As an alternative, the table can be underlit to allow the reproduction of transparencies, which can be any size from about half that of a slide up to about 25 cm × 20 cm (10 in × 8 in). The illuminated stage for transparencies is lit by a special three-color lighting unit which is used both for adding color to black and white originals and for the color matching or correction of color transparencies.

With a simple table the camera operator must position the item to be reproduced. When, as is often the case, precision positioning is needed, the artwork or transparency is placed on an animation compound. This device allows the original to be moved under the camera by turning handles, one for the X-axis (east-west), one for the Y-axis (north-south) and one to rotate the original through 360 degrees. Each movement is precise to a fraction of a millimeter and can be read off from a display counter. Repeat accuracy is essential.

The camera itself rides on a rigid vertical column that may be 3 m (10 ft) high or more. Although manual focusing is available, the camera usually has an automatic focusing arrangement that ensures that the lens focus adjustment is moved exactly as required as the camera rides up and down the column. This will be located at the top of the column when photographing large pieces of flat art, and at the bottom when copying from another slide.

The camera itself can also be changed. The standard camera movement is for 35 mm slides, but this can be exchanged for a camera designed to take superslides on 46 mm film, or even for a 16 mm camera for filmstrip or movie making. Whichever film is being used, the camera accepts bulk-loaded film in lengths of up to 120 m (400 ft). The main feature that sets the camera apart from a normal 35 mm camera is pin registration.

Figure 2.11 A Forox rostrum camera in use. The artwork or transparency to be copied is carried on an animation compound which allows precise location and planned movement. (Photo Prater Audio Visual)

Many of the effects that can be achieved by the rostrum camera depend on the precise positioning of the image relative to the film. For example, one of the most common techniques used in slide making is double exposure. A number of frames may be exposed for one piece of text, and then some of them exposed again to add further text as part of a build-up. If each exposure is not perfectly in register, the result will be a mess. Thus, the camera is equipped with registration pins that precisely locate the film in the camera for each exposure.

The camera also has controls that allow for multiple exposure, automatic taking of a preset number of exposures, and automatic film movement both forwards and backwards. Camera movement up and down the column is motorized, and the whole ensemble can be under either manual or computer control. Viewfinding can be done either with an eyepiece viewfinder, or by a system called reticle projection, where an image corresponding to the exact slide format is projected back through the camera on to the artwork or slide stage. This allows the artwork and camera to be precisely positioned so that only the desired material is copied.

What can a rostrum camera do?

A rostrum camera can copy any piece of artwork, photograph and color transparency on to a slide. It can select any part of the image, or can make up a series of slides from one original (e.g. for multi-image work). It can also do several less obvious things.

It can add color. Most professionally made text slides start life as a typeset item in black and white. This will then usually be transferred to high-contrast black and white film known as ortho film (e.g. Kodalith™). For some jobs this slide may be used directly or else be sandwiched with a colored gel; but the more usual requirement is for colored lettering on a different colored background. This is done by exposing a positive (black lettering on clear background) version of the slide, illuminated by the required background color. The same piece of color film is exposed again, this time to a negative (clear lettering on a dense black background) version of the slide illuminated by the required letter color. The final result is a two-colour slide.

Figure 2.12 Those users not needing the productivity of the full-size rostrum camera, but still needing many of the effects, can use a simpler copy stand with a pin registered camera. (Photo of the MRX Camera from Image Ltd)

This process shows why registration is vital. Any discrepancy between the positive and negative images would show up badly. The process is not limited to two colors; further multiple exposures will allow as many different colors as required. In the case of text and graphics slides an extension of the process allows the production of a series of slides constituting a build-up, each component of which is in precise registration with its predecessor.

The rostrum camera can be used to produce many trick effects in the hands of a skilled operator. These include:

☐ Streaking, achieved by a combination of many successive exposures and zooming of the camera lens.

☐ Neon glow, achieved by several methods including placing a diffuser glass a short distance in front of the text being copied.

☐ Spinning text, achieved by multiple exposure and rotation of the object text.

☐ Masking the image with a photographically produced mask of any shape. The mask can be hard or soft edge. By extension of this technique photo montages can be made up, with images merging into each other or occupying defined areas.

☐ Posterization, and other tricks with color. An image may be rephotographed to black and white through different color filters, and then reconstituted as a color picture with completely different colors.

The modern rostrum camera is a highly efficient means of producing visuals, with enormous scope for creative use. Not everyone needs all the facilities of the big rostrum camera, so users with less exacting requirements use simpler copy stands (see Figure 2.12) with cameras that have the vital attribute of pin registration even if they do not have a big film capacity.

In-house or out?

Most OHP transparencies are made at short notice within the organization that is going to use them, but most slides are made by specialist service companies. The exception is the large organization able to justify its own fully equipped AV department.

The description of the rostrum camera makes it clear why this is so. A skilled camera operator is able to produce high-quality slides in large quantities at a reasonable price. Most service companies quote standard prices for the commonly required slide formats, and will always pre-quote for special work.

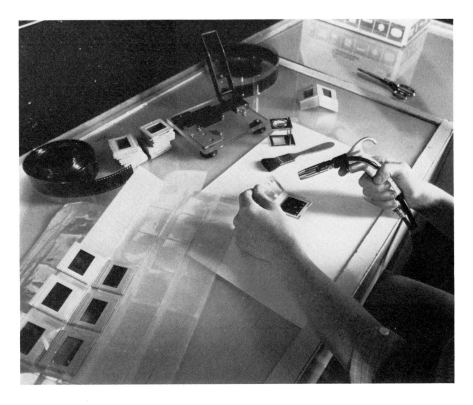

Figure 2.13 Care must be taken when mounting slides to ensure that no dust is trapped in the mount. (Photo Tony Gidley Productions Ltd)

Slides can be professionally made at short notice. The service companies typically offer a three day standard turnround, and 24-hour turnround at a premium. Even so, last-minute slides may be required. This need can now be met in-house by the use of instant slides. Film for instant slides is available from Polaroid in 35 mm cassettes, which can be used in any normal 35 mm camera.

Slides can also be created by a computer, using a suitable graphics program. This method of producing visuals is now becoming so important that it warrants a separate chapter of its own. Although computer-generated images are expected to take an increasing share of an increasing market, there will remain many types of image for which the photographic method is either the only or the best way. Most slide service bureaux will probably offer both types of slide production in the future, and will match the method to the client's requirements.

Chapter 3

The role of computer graphics

If this chapter had been written in 1980 it would have been of limited interest to the average AV user, because the relevant computer graphics equipment available at the time was extremely expensive and able to serve only a limited market. Now the situation is completely different, and anyone who has to commission visuals must be aware of what computer graphics can do. The problem is that the computer graphics industry has enjoyed explosive growth, and there are now hundreds of companies offering equipment and services. It can be difficult to determine why apparently similar products can vary so much in price and performance. However, the situation will probably settle down soon, and in each market there will be clearly defined sectors served by a relatively small group of suppliers.

The basics

A brief description of how computer graphics work will help put the subject into perspective, and explain why there has been the sudden take-off in activity. One of the components necessary for practical graphics is *memory*, huge quantities of it. It is the amazing drop in price and

increase in density of semiconductor memories (and similar components) that has made high-quality computer graphics available to all.

The way in which computer graphics are seen is, first of all, on a cathode-ray tube – the display component of all TV sets and video monitors. An electron beam is deflected by electrodes so that it 'writes' on the tube surface; when the beam hits the phosphor on the inside of the tube, the phosphor fluoresces. The beam can be modulated so that the resulting spot of light varies in brightness.

An image can be built up on the tube face in one of two ways. In computer graphics they are referred to as vector and raster graphics. Vector graphics can be considered as a set of instructions describing how to draw a particular picture. If the device drawing the picture can think in terms of coordinates on the screen, it can store the instruction set in the form of:

☐ Start at 100x, 100y
☐ Move to 245x, 100y
☐ Move to 100x, 320y
☐ Move to 100x, 100y

This succession of moves draws a triangle. If the movement or 'vector scanning' is done fast enough, and the phosphor of the tube has some persistence, the triangle is seen as a still image. The process of continually redrawing the picture is referred to as refreshing the image, and is usually done about 30 times a second.

A system for vector graphics therefore consists of a means of entering the coordinate information, and a memory to store the long sequence of instructions needed to draw the image. Vector graphics can be extremely precise, the accuracy is limited by how precisely the coordinates are specified. The graphics can include arc drawing routines, so that circles and curves look smooth. Vector graphics are widely used in computer-

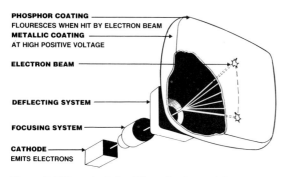

PHOSPHOR COATING
FLUORESCES WHEN HIT BY ELECTRON BEAM
METALLIC COATING
AT HIGH POSITIVE VOLTAGE

ELECTRON BEAM

DEFLECTING SYSTEM

FOCUSING SYSTEM

CATHODE
EMITS ELECTRONS

Figure 3.1 The principle of the cathode-ray tube.

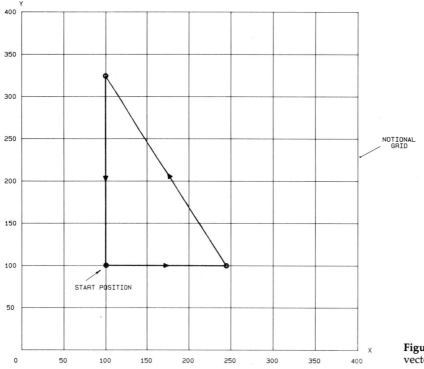

Figure 3.2 Drawing a triangle with vector graphics.

aided design and drafting, where the accuracy of the line drawing is essential, and where mechanical plotters can produce a hard-copy drawing which follows exactly the same vector instructions used to compose the picture on the cathode-ray tube.

Raster graphics

Although vector graphics are accurate, and are excellent for applications like architectural and engineering design, they are not so flexible when large masses of color are required. There is also a problem in displaying vector graphics on a standard video monitor or TV set.

In a monochrome display of a standard television set the cathode-ray tube is used in a different way than in vector display. Instead of moving in a way determined by the image; the electron beam moves in exactly the same way whatever image is to be displayed. It scans the tube face from the top to the bottom, building up a raster of lines.

In the USA the image usually consists of 525 lines, with a complete image being scanned 30 times a second. In Europe the image usually

Figure 3.3 Computer graphics are widely used in engineering design and drafting. Many of the block diagrams and flowcharts in this book were prepared in the drawing office of Electrosonic Ltd using a CAD program running on an IBM AT computer linked to a Hewlett Packard pen plotter.

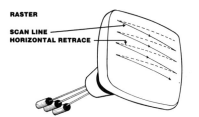

Figure 3.4 Raster display by a cathode-ray tube.

consists of 625 lines, and a complete image is scanned 25 times a second. In fact, most TV images are interlaced with, for example, two fields of about 312 lines. One field takes one-fiftieth of a second to scan, and the next field places its lines between those of the first field.

Whether an image is seen or not depends on the modulation of the electron beam current. No beam current means no picture; maximum beam current causes a bright line to be drawn.

A color tube is just three times as complicated. Subjectively, any color can be built up by mixing the appropriate proportions of red, green and blue. The face of a color TV tube consists of thousands of phosphor dots arranged in sets of three; one dot in each set fluoresces red, one blue and one green. Usually three electron guns are used in the tube to produce three electron beams, one for each color. A clever masking arrangement ensures that the red gun can only hit the red dots, and likewise for the blue and green guns.

The resolution of a particular color display will therefore depend on two factors: how finely the dot pattern is printed, and the quality of the electrical signal controlling the electron beams. The normal TV picture is made up from an analogue signal. This means that the electrical voltage that modulates the beam current moves *continuously* over a range of values to change the momentary brightness of the picture. The picture quality will, in principle, depend on how fast the

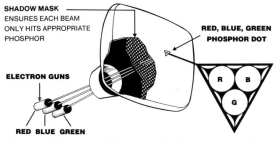

Figure 3.5 The color TV tube. In this illustration the dot pattern shown uses round dots, but some tubes use rectangular or stripe dots.

signal can change values, and on the precision with which a particular value can be maintained. While this method of image construction is very satisfactory for live picture sequences created by a video camera (whose internal action is a scanning action which neatly matches that of the cathode-ray tube display), it does not lend itself to computer synthesis.

The principle of raster graphics is to notionally divide the screen into a large number of picture elements, abbreviated *pixels*. An obvious possibility is to match the capability of the graphics system to that of a TV image. In a 625-line system not all the lines are displayed, so a popular raster display is based on a pixel array of 576 (vertical) × 768 (horizontal), which means that each of 576 lines is divided into 768 pieces.

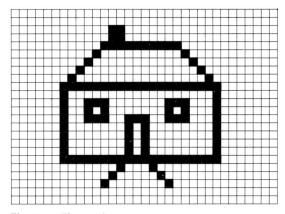

Figure 3.6 The pixel concept.

A computer works entirely on a digital basis: that is, it can only think in terms of discrete values. It might seem reasonable to ascribe to each pixel a range of values for each of the red, blue and green components. Reasonable, yes, but quite demanding on the computer memory because in the example given values for 442,368 pixels would need to be stored.

Computers store numbers as binary digits or *bits*, each of which can only have the value 0 or 1 (on or off). A convenient grouping is a *byte*, which consists eight bits and is thus able to represent 256 different values. A typical store for a raster graphics display allows one byte per pixel, and it becomes a matter of user choice as to how the values are ascribed.

In theory, the user could choose from a range of options, for example:

☐ 256 different colors, each created by a preset mix of red, green and blue;

□ red, green and blue;

□ 256 different levels of white;

□ 8 levels of red, 8 levels of green and 4 levels of blue.

In practice the first option is the one often used, and the user is able to define which colors he will use.

It is already clear that the quality of raster graphics is limited in two ways. The *resolution* is limited by the number of pixels in the display, and the *color* and *intensity* precision are limited by the number of memory bits allocated to each pixel. Both these points are vitally important for some applications, and are discussed later in the chapter.

A typical system

A typical computer graphics system will consist of:

□ A display, which is usually a color monitor. Depending on the precision of the work to be undertaken this can be either an ordinary commercial monitor or a precision device with high resolution, which is much more expensive.

□ A frame store. This stores the instantaneous values of every pixel in the image being presented.

□ A picture generator. This converts the information held in the store to the separate red, green and blue signals needed by the display.

□ A central processor (or computer) that controls the whole system and, in particular, allows the user to easily change what is in the memory.

□ A keyboard, for entering instructions and text.

□ A long-term memory, usually a floppy disk or hard disk store for the long-term storage of completed images.

□ Sometimes a second monochrome monitor for displaying instructions, program options etc.

In reality, a complete computer graphics system consists either of a purpose-built system or, increasingly, a standard computer to which a graphics package has been added. The graphics package consists of a special graphics controller board that carries out the picture memory and picture generation functions, together with computer software allowing easy access to it. Graphics packages are available for popular computers such as the Apple™ and IBM PC™.

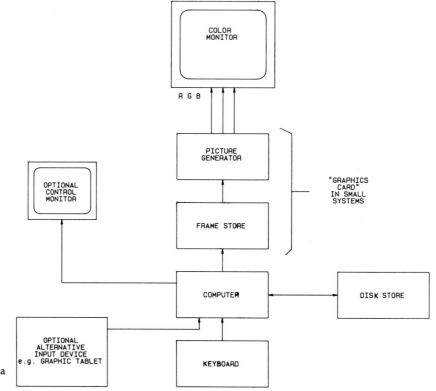

Figure 3.7 Block diagram of a computer graphics system.

What can such a system be expected to do? In the past there tended to be a distinction between graphics programs based on the setting-out of text and those intended for artwork, sometimes referred to as paintbox programs. Customer demand and technological advance have now brought both together, so a reasonable power graphics program would be expected to allow:

☐ The setting-out of text anywhere on the screen. Early programs only allowed the use of lettering used in normal data displays. This is virtually useless for display and presentation work. Therefore, the program should allow a reasonable choice of fonts and character sizes. It should be possible to position text anywhere on the screen.

☐ The selection of any required color. Computer graphics people advertize somewhat meaningless claims such as 'a palette of 16.7 million colors'. In practice, a more realistic choice will be given at any one time – for example a choice of 256, but with the possibility that the user can re-define a color. There is sometimes a restriction on the number of different colors that can be used in a single image, e.g. sixteen.

☐ The drawing of any outline. This can be on a freehand or a construction basis. In the case of construction the user defines the beginning and end of lines and the computer draws the straight line between, or the center and radius of a circle are defined and the computer draws the circle. Here the computer is exercising the principles of vector graphics but translating them into raster form.

☐ The filling of a shape with color. This is picture-book painting made easy. Any outline which is defined by an unbroken line can be filled with a selected color.

☐ The storing and manipulation of a shape, image or pattern. For example, a company logo can be lifted from a back-up memory and placed anywhere on the screen. It can be enlarged or reduced and multiple copies can be made.

Image entry

In practice each program will have other special features – for example the ability to modify lettering by rotation, tilting or adding drop shadow, or the ability to draw or shade pictures using a selection of 'brushes' with fine or broad brush strokes. An 'airbrush' facility is often included. It would be impractical for the user or artist to enter all painting instructions by a keyboard, so the keyboard is used for text entry and system instruction and everything else is entered using a graphics tablet, 'mouse' or other special input device.

The graphics tablet is the most commonly used. It contains a coordinate grid so that, if a stylus or cursor is placed at a particular point, the position of that point can be recorded. At first it may seem difficult to both look at the monitor and draw on the graphics tablet, but artists quickly get used to looking only at the monitor and their painting hand follows the activity on the screen. They only look at the graphics tablet if, for example, an overlay is being used and an existing image, such as a logo, is being digitized for subsequent manipulation.

Figure 3.8 A professional computer graphics design station. (Photo Dicomed Corporation)

Digitizing an existing image by hand is tedious. Again electronics comes to the rescue. A video camera can be used for 'frame grabbing'. Any piece of existing artwork can be viewed by a video camera and transferred into the appropriate memorized pattern of pixels. Often only a simple monochrome camera is used, and color is added by the artist once the frame has been grabbed.

Full-color frame grabbing is also possible using a color camera. The problem here is that the original video image will usually have far more color information in it than the graphics system can handle. Either some arbitrary color selection must be made, or the frame-store memory greatly increased, usually to a 24 bit per pixel system which allows 256 levels of each of red, green and blue. This gives an image which, in color terms, is indistinguishable from the original.

Once an image is in a computer memory it can be modified in many ways. Because the entire image is recorded in numerical form, it is easy to apply mathematics to some or all of the numbers. In simple graphics systems this may have the effect of changing a color. In the advanced graphics systems used, for example, in the processing of images received from satellite cameras or thermal imaging cameras, the computer can be used for image enhancement. This process of making the invisible visible depends on the detection of very small changes in the nature of an image which are then amplified.

Image output

The foregoing brief description of how an image is assembled in a computer is relatively easy to understand. In practice graphics programs vary enormously in complexity, and those at the top of the range employ considerable sophistication in providing routines that speed up the image creation process. Some programs are application specific. For example, business graphics programs usually include automatic chart and graph drawing routines, where it is only necessary to feed in the numerical values to be represented.

Having created the image on the computer screen, what next? Sometimes it is sufficient to keep the image in the computer, in which case there must be some easy way of storing and retrieving images. This possibility is discussed further in Chapter 20. More often, the user requires the image in another form.

The business user will be using the output to support some kind of presentation. For a written presentation, a printed version of the graphic can be provided by a suitable plotter, either in monochrome or color. This is also satisfactory for simple graph and chart work, but because the plotter is usually plotting an exact replica of the screen memory, the pixel structure is obvious. Furthermore, this type of output can only provide a limited choice of color.

If an overhead transparency is required as the final output it can be easily made on suitable plotters, or can be reproduced from the hard copy. It is also possible to make transparencies photographically, direct from the computer screen.

What is the best way to get the output in photographic form? The obvious method is simply to point a camera at the color monitor. If the correct shutter speed and suitable film are used, the results are remarkably good. This is now made very easy with low-cost cameras using instant film, although ordinary single-lens reflex (SLR) cameras are also quite suitable. A hood arrangement ensures that the screen image is not affected by ambient light and that the camera is exactly the right distance from the screen.

While this method is appropriate for the production of business graphics slides for in-house use, such as a weekly or monthly management meeting, the results are not really good enough for professional presentation or for print origination. The first major limitation is the color monitor tube, which introduces both distortion and a textured look to the image arising from the tube's three-colour dot pattern.

This can be avoided by using a specially designed film recorder. These devices vary greatly in complexity, but in principle they consist of a high-definition *black and white* monitor with a flat face plate. The camera, which can range from a simple SLR camera to a sophisticated camera with pin registration, is in a fixed position directed at the screen. In front of the camera lens there is a color filter wheel, which can successively interpose a red, green and blue filter between the screen and the camera lens. The whole assembly is in a light-tight box.

The film recorder receives the image in the form of its separate red, green and blue components. It displays each in turn, for the appropriate time, as a precision black and white image. Because of the insertion of the associated color filters, the camera records the full color image.

Slides produced in this way are of high quality and have excellent color. However, because of the high quality the defects of the image creation process are emphasized, and the pixel construction of the image is extremely obvious when blown up by slide projection. Circles and diagonal

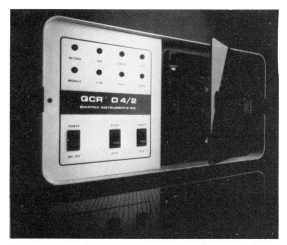

Figure 3.9 A professional film recorder. These are available with different resolutions, typically from 1000 line to 4000 line. (Photograph Matrix Instruments Inc.)

lines suffer from the 'jaggies', which can be acceptable for informal presentation work, but are not acceptable for professional use.

Image enhancement comes to the rescue. Because the image is stored in numerical form, it can be processed to yield a higher-definition picture. The main method is to greatly increase the number of lines in the image in order to eliminate the obvious raster construction, and to increase the number of pixels in the horizontal line. Medium-resolution systems offer between 1300 and 2000 lines, high-definition systems between 4000 and 8000 lines.

Needless to say, the high-definition systems are expensive. The business user planning to use computer-generated high-definition graphics has a range of choices.

The choice

Choosing the method of making computer-generated slides (or OHP transparencies, artwork etc.) depends both on personal inclination and on objectively answering a number of questions:

☐ How many slides are required annually?

☐ It is a *requirement* that slides be made in-house (e.g. for security reasons)?

☐ What level of quality is required?

☐ What sophistication is required in the preparation of artwork (as opposed to text only)?

☐ Who is going to operate the system?

This last question is probably the most important. If all that is required are simple text and chart slides, anyone with rudimentary keyboard skills can produce acceptable results. However, as soon as any element of design or creativity is required, good results are only obtained when an artist or person with graphics training is at the controls.

In ascending order of cost the choices are:

☐ To use the low-definition system based on a personal computer and make slides directly from the screen. This is only suitable for non-critical once-only work for small meetings.

☐ To buy a computer program, such as Dicomed's Presenter™, which allows the creation of slides on a low-cost computer like the IBM PC™. The data held in the computer is then transmitted to a service bureau, either directly by telephone line, or in the form of a floppy disk; The bureau makes a high-definition slide for about $10 a shot. The results are excellent, and the bureau can usually help enhance slides with additional artwork, such as logos, or can produce the more difficult material as an additional service.

☐ To buy a complete stand-alone system. If, for reasons of security, it is essential to have the entire system in house, there is no choice but to have a complete package including the film recorder. Systems using standard SLR cameras vary in price from around $12,000 to $80,000. The price difference is accounted for mainly in definition, which ranges from around 1000 lines to 4000 lines, and in the sophistication of the graphics software.

☐ If very high definition is required, and a large throughput is expected, then a top of the line system will be warranted. These cost $250,000 and upwards, and are used either by bureaux or by very large users.

Only the first three choices are of interest to the business user. In Europe there is now a much

Figure 3.10 By using a program such as Dicomed's Presenter™ it is possible to produce high-quality slides on a personal computer.

Figure 3.11 Typical of the complete stand-alone systems is the Starburst graphics slide creation system. It is widely used by both end users and bureaux. (Photo Mediatech Ltd)

greater tendency to use the bureau approach; usually the slide-making bureau is asked to do the whole job, but sometimes the initial creation is done in-house. In the USA more complete systems are purchased by end users, although whether the systems are used enough to make them a worthwhile investment is open to doubt.

In all the enthusiasm for computer-generated material, it must not be forgotten that there are many cases where traditional all-photographic methods produce better results at lower cost. Thus most slide-making bureaux are beginning to offer a combination of resources, so that the end user does not have to worry about the actual method used.

Formats

One cautionary word: slides are correctly presented in the 3:2 format and video images in the 4:3 format. Professional computer graphics slide-making systems have as their end product a properly proportioned slide. Low-cost systems sometimes limit the user to a slide in the video format, which may be unacceptable if it is to be used with other photographically produced material.

Video output

Can a computer graphics display be directly transferred to video tape? It might seem easy, because the original image is shown on a color monitor. However, in computer graphics the

Figure 3.12 The Spaceward computer graphics system has been speciallydeveloped with the needs of television in mind. It can produce complex graphics sequences at remarkably low cost.

image is stored in its component red, green and blue parts. A normal video system needs a composite image. Some slide-making systems are able to offer a composite (NTSC or PAL) output, and these images can be transferred one at a time onto video tape, using a suitable editing system.

More usually, a graphics system designed specifically for video output will be used. These work within the limitations of the standard video image, and make no attempt at a high-definition output. They do, however, make much more of the image sequencing possibilities, and ultimately allow complete animated sequences to be produced.

These sequences are normally produced in slow motion, because the task of replacing one complete image with another involves the movement of a massive amount of information. If this has to be done 30 times a second, a very high speed and large capacity computer will be needed.

Chapter 4

Making audio-visual shows

The compartmentalization of AV into film, video and slide/sound (or multi-image) has given the unfortunate impression that these are completely separate media with little in common. But, in the same way that the rules for making visuals for visual aids or programs are exactly the same whether a slide, overhead transparency or video page is being made, so the rules for making an AV program, whether a film, videotape or multi-image show, are also the same. They all require discipline.

Film was the first medium to impose a discipline, because of the cost of film stock and the fact that the results could not be seen instantly. On the face of it, video and slide/sound are easier because everything is so accessible. This very accessibility resulted in some truly dreadful productions, until users realized that these media also need a systematic approach. Not surprisingly, the techniques of video and slide/sound have moved towards the methods of film making.

The procedure for making an AV program, regardless of medium, is summarized in Table 4.1 and discussed in the next few sections.

Define objectives

Even if the objectives of an AV production have already been defined in a briefing document, the first stage of any production must be to thoroughly examine why it is being made in the first place. The sponsor and producer must be able to clearly define:

□ The benefit(s) that will accrue to the sponsor as a result of the show being made.

□ The action that is expected from the audience as a result of seeing the show.

It is best not to start by saying that a certain piece of information must be communicated. The danger of doing this is that the obvious route will be taken, which usually results in far too much being put into the program. By being absolutely clear about objectives, only those items that are relevant will be included.

For industrial, commercial and training AV shows there is a very simple rule: *one show, one message*. The temptation is always to try and do too much. The process of defining objectives may well reveal that there are multiple objectives that are best dealt with separately. Sponsors should not be worried about how short a program is, because it is generally better to make two or more short programs than one long one. Often the group of programs can be treated as a single project, at no significant extra cost.

Table 4.1 Making an audio-visual show

1. Define objectives
2. Choose medium
3. Agree budget
4. Prepare first treatment
5. Confirm costs
6. Write script and prepare storyboard
7. Carry out location photography and prepare special artwork
8. Carry out rostrum photography
9. Prepare sound track
10. Edit or program
11. Preview
12. Correct if necessary
13. Make show copies
14. Show
15. Evaluate

Choose medium

The choice of medium must be made on the basis of the considerations discussed in Chapter 1. The definition of objectives will reveal the kind of message that is to be put across, which will determine, for example, whether any moving-picture sequences are required. More importantly, this part of the production process will thoroughly analyze how the final program will be shown. For example, whether it will only be shown to group audiences, or to individuals; whether it is for a fixed single intallation, or should be made in multiple copies; and whether it is for a single special occasion, or for continuous daily use over several months or years.

Often a hybrid solution is favored. A multi-image show may be ideal for the big audience expected at the annual sales meeting, but the same show may be required in a portable version for later use. In principle, it can easily be transferred to video, but *only* if this is planned in advance, because this will impose restrictions on the format and content of the original multi-image version.

Agree budget

It may seem an obvious statement, but it is amazing how often sponsors of AV programs do not take account of it. The objectives of an AV program should determine the budget.

For many programs this simple statement can be literally applied. If the audience respond in the required manner (by buying the product, by reducing absenteeism, by eliminating waste) a direct cash benefit can be seen. The question then simply becomes one of how much of this benefit can, or should, be applied to the show budget. If the planned show is going to cost more than the benefit, it is doubtful whether it should be made.

It is more difficult, but still possible, to apply the same criterion to motivational, public relations, and public entertainment shows. Often it can be a matter of dollars per head. Sometimes a show is made to make a small group of people, or even one person, feel good. A senior manager may want a program made that makes the chairman of the board look good, and to keep the chairman off the back of his subordinates. If he is a good manager, he is not only honest with himself and knows that this is the *real* objective of the show, but can actually ascribe a value to it (e.g. the cost of recruitment and disruption if he loses some of his subordinates).

It is important to include in show budgeting the realistic cost of delivering the show to the intended audience. It is no good making a superb show which, in practice, costs thousands of unbudgeted dollars to stage each time it is shown. It is equally no good making a video for use by sales people if the cost of equipping each member of the sales force with a video set has not been taken into account.

Prepare first treatment

This is the stage at which the show producer prepares an outline of the show. In the case of a major production this stage includes a draft script and storyboard, but for a small industrial AV, it is simply a summary of how the objectives are going to be met. The summary need only be worked out in sufficient detail to quantify the resources needed for the production and to provide a production timetable. This may also serve as a 'client-approval point' when an outside producer is making a shown for a customer.

Confirm likely costs

The work done for the first treatment must be sufficient to allow all likely costs to be confirmed, and to ensure that the show is going to be delivered within budget. The rule is that if the costs are going to exceed budget, the first treatment must be done again and again until the two match.

Rules are sometimes made to be broken. It can be that the first treatment itself will reveal further benefits that could result from a more expansive approach. The show sponsor must decide if the extra benefit is worthwhile. If not, he must insist that the show sticks to budget.

Write script and prepare storyboard

This is the vital creative stage of production. Much of the succeeding work can be done by individual experts, all of whom can make a contribution to the quality of the show. What they cannot do is change the overall approach or basic script.

Scriptwriting is an underestimated task and a most valuable skill required in any AV production. A good script gives the other creative participants the best opportunity to shine, and, of course, is the key to the show meeting its objectives.

But the script cannot be written in isolation. The show storyboard must be prepared at the same time. A storyboard is a series of sketches showing the visual development of the program. The principle is the same for a feature film or a five-minute training program. A feature film will have an elaborate storyboard that looks something like a strip cartoon, and will even include precise directions for camera angles and likely final editing. The storyboard for a simple production need only consist of rough sketches or a series of written descriptions of the visuals that must be obtained.

There are no set rules as to how a storyboard is presented. Some producers use a looseleaf book with pre-printed pages, each page having, say, four frame outlines which can be sketched in, and space next to them for the corresponding script together with a description of sound and visual effects required. Others use a system of pre-printed cards, with each card representing a key visual. This arrangement is more flexible because it allows extra cards to be inserted as the storyboard develops.

Writing AV scripts is *not* the same as writing a radio script or a piece of text. A radio script, for example, is written so that listeners are forced to use their imagination. It must include enough clues to allow listeners to visualize a place, person or object. In an AV script the visuals are used to convey a *precise* visual impression. The script must not describe in detail something which should, in any case, be clear to the audience. Thus the process of writing an AV script tends to be a two-part one: first a draft script, followed by heavy revision and cutting in the light of the storyboard development.

Although there are some valid uses of AV technique that rely on long verbal explanations of a single still picture (e.g. in maintenance instructions for complex equipment, or when AV is used to give assembly instructions in manufacturing) most AV shows must flow. In general, they will have a new principal image every six seconds or so. This applies whether it is a moving picture (the new image being a shot from a different angle or a cutaway to a different image) or a series of stills. Some motivational shows use a much higher picture rate; multi-image sequences using one slide every two seconds are quite common.

This reveals another truism: the complexity of a message is inversely proportional to the complexity of the means of its delivery. An educational slide/sound sequence, intended for an audience of one person at a time, may be used to explain a complex piece of organic chemistry, but needs

Figure 4.1 The AV producer works from a storyboard. The show is planned in great detail. (Photo Tony Gidley Productions Ltd)

only a straightforward factual script supported by clear visuals. On the other hand, a show that is designed to motivate a large audience and make them feel they belong to one happy family may require considerable resources, combining movie with multi-image and calling for the ultimate in the scriptwriter's art. In the words of one AV scriptwriter, 'emotion costs money'.

The completion of the script and storyboard is another approval point. In fact, it is the last before the show is completed. From now on the sponsor should keep out of the way and let the producer get on with the job. If the sponsor has not got confidence that the producer can make the show, the producer should not have been appointed in the first place!

Location photography and special artwork

Now the actual process of production can start. Some shows are made entirely from existing material, others consist entirely of original photography. Most are in between. At this stage the producer must:

☐ Assemble existing material. This could include existing slides or artwork, and existing film or video material.

☐ Prepare special artwork. This ranges from the complexity of animated cartoon art, to the more usual requirement of individual pieces of art or typesetting. The choice of using conventional or computer methods will be determined by the style of the production.

☐ Obtain original photography. Depending on the medium chosen this will be still, film or video movie photography. A shooting schedule must be prepared showing exactly what is required, and where and when it will be shot. Much AV work is based on short studio sessions, but outside location photography may also be required.

☐ Obtain location sound. Any 'lip sync' sound forming part of a drama-style production will normally be recorded at the time of filming. Likewise any unique sound effects that could not be created in the sound studio later must be recorded when the corresponding photography is done.

Rostrum photography

The use of a rostrum camera to prepare visuals has already been described in Chapter 2. Rostrum photography is an essential part of the production of professionally produced slide and multi-image shows. It is also often a major component of film and video sequences.

Some shows are made using multi-image techniques which are then transferred to video or film, using rostrum cameras fitted with movie or video camera heads. Besides allowing the usual single-image visual effects, they allow the construction of image sequences where an original image is panned across, or zoomed into. Many documentary films are based on the use of still images where movement is imparted by the use of the camera and compound table movements. An extreme example of the rostrum camera is the animation stand, used for making animated cartoons. Although this type of work can, in theory, be done on a simple rostrum camera, in practice the demands of animation are more complex and require, for example, the ability to shoot backgrounds separately from the foreground figures.

Rostrum cameras used for slide-making can benefit from computer control, but it is by no means essential. However, rostrum cameras used for the direct production of video sequences require some level of computer control because of the technical difficulty of laying down video images frame by frame. When rostrum cameras are used for movie film production there is also a choice, biased in favor of computer control.

If direct computer-generated animation is required for a production, the origination also takes place at this stage.

Sound track

Some elements of the sound track may already have been prepared as part of the original photography, but further original material can be assembled in the studio. At this stage, the procedures for slide/sound or multi-image programs may diverge from those of a moving picture program.

A slide, filmstrip or multi-image program usually involves making a sound track to serve as the final track, including narration, music and effects. The images are then matched to the sound track during the programming stage of production. With movie and video, the procedure tends to be the other way round; the sounds are edited to fit the edited visual sequence. Thus, while live sound will simply be edited together with the pictures, music and narration may not be added until the main visual sequence is complete.

An exception may occur when a specially composed music sequence is used, or where a tight narration (as in a commercial) must fit a precise time slot. In this case, the sound recording

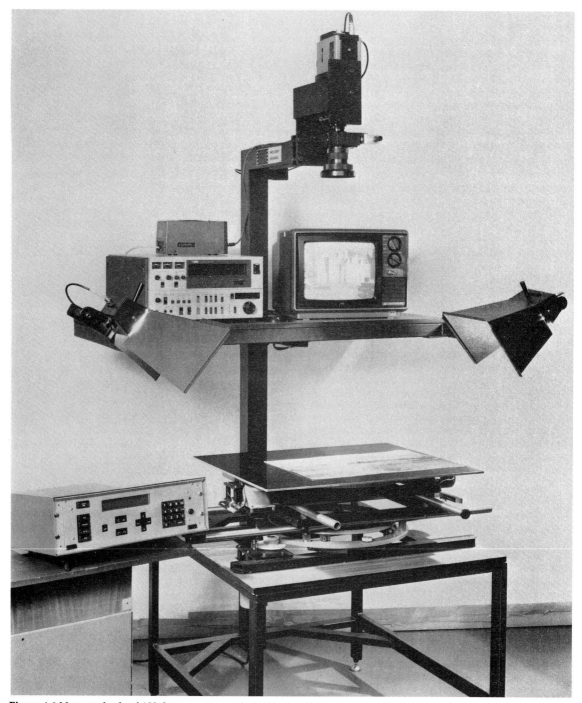

Figure 4.2 New methods of AV show creation include the use of a computer-controlled video rostrum camera. This system can transfer single images or programmed sequences direct to tape. (Photo Neilson Hordell Ltd)

is made first, and the visual editor is left to solve the problem of making the iamges exactly match the words or music.

The problems of synchronizing a lot of different source material have been greatly simplified by the introduction of time codes, referred to in more detail below and in Chapter 16.

Edit or program

This is where the show is finally put together. In a slide or multi-image show the sound tape is now complete and all the slides are ready for loading into slide trays. The programming of a simple slide/sound sequence may simply involve loading the slides in the required order, listening to the tape, and pressing a button at the moments when a slide change is required. Changes are made by redoing the whole or a part of the programming.

A multi-image show is more complex. Although the principle is the same, there is now too much material to do the programming in 'real time', so the process of loading the slides tends to be combined with the programming and previewing sections of the show; see Chapter 16.

This allows changes to the timing of visual effects to be tried out and very precise synchronization to be achieved.

Film editing is an extremely disciplined process. Normally the film editor will do all his initial work on a work print of the film. Editing is time-consuming, but is not capital intensive, because the editor works at an editing table that allows the print to be viewed on a small screen. As part of the editing process decisions will be made about how the transition between one shot and another will be made, whether as a 'cut' or a 'dissolve'. The actual effects can only be achieved at the time the final print is made. The film laboratory cuts the negative exactly as the editor cuts the work print, but into two reels, the A and B rolls, which consist of alternate pictures and black. The optical printing machine is programmed to dissolve as required between the two overlapping pieces of film.

Traditionally, sound was edited by hand as well. Each sound element is recorded on to magnetic film, which is sprocketed just like the picture film. The film editor cuts each sound effect to match the film, and the resulting sound elements are dubbed on to a master magnetic film which has exactly the same number of frames as the corresponding picture film.

Video picture editing is heavily dependent on electronics. It is not practical to cut and splice

videotape, although this was actually done in the early days of video recording. The reason is that video editing consists of linking electrical signals, and unless this is done at precisely the right moment, the result is a bad picture roll. Therefore, the technique now used is electronic editing. In principle, each piece of video material is copied across onto a master tape and the effect of switching-in or fading-in a new image is under precise electronic control.

Early video editing systems suffered from a lack of precision. Video engineers soon realized that their problem was that they had nothing equivalent to the discipline imposed by the sprocket on movie film. With movie film everything could be calculated in advance in terms of the number of picture frames. Great precision could be achieved in the synchronization of picture and sound effects because each element could be precisely locked together.

The solution to the problem is the time code. Here a separate track on the videotape is used to uniquely identify each frame of video material. The same code can be used on audio tapes. This allows for rapid location of material and for the ability to run different media in 'frame sync'. It also keeps costs down.

High-quality video equipment is extremely expensive. A broadcast-standard editing suite can cost several hundred dollars an hour to use. The editor no longer has the luxury of a lot of time to make editing decisions, so editing may be rushed. The answer to the problem is to time code all the original material and then copy it onto a low-cost video format (U-format or even VHS). The editor can then make all basic editing decisions 'off line' by noting all the time code readings of the edit points – working, in fact, in much the same way as a traditional film editor. In the same way that the traditional editor got the film laboratory to do the final negative cutting and printing, the off-line video editor goes back to the big edit suite to have the final tape made up from the original material, but does not waste any time taking edit decisions.

Any video special effects are added at this stage, and the sound track may also be completed. But maybe not; because the flexibility of the time code system allows the original sound track to be made in a separate audio studio if appropriate.

Not surprisingly, the use of time code has infiltrated the traditional movie world, allowing high-quality sound studios to prepare sound tracks on conventional (non-perforated) tape which can be electronically locked to the movie. Traditionally, final film dubbing is done in a dubbing theater where a work print of the final film is shown, and the sound track is assembled on interlocked magnetic film transports. A new trend is to make a video copy of the film with the same time code, and to make the sound track in a conventional audio studio with a video player that can be time-code locked to the master audio recorder. There is thus a great coming together of the audio-visual media, with a wide choice of methods of achieving the finished result.

Preview

Now for the moment of truth. While someone previewing their own work may be happy to see the final show under any circumstances, those producing a show for a customer must ensure that the circumstances of the preview allow the proper presentation of the finished product. It is essential that it be seen in a proper viewing theater, or, in the case of a complex multi-image production, in an environment that approximates to that in

which the final show will be run. On no account should the client be allowed to feel that the production is anything other than complete.

Correct if necessary

If the production has been properly planned the preview should result in congratulations all around. Changes should not be needed, because any changes required should have been spotted and made at the editing/programming stage. Changes made at this stage are expensive in all but the simplest shows, but, if unavoidable, mean an extension to the editing process.

Make show copies

The show seen at the preview stage may represent the one and only needed 'show copy'. This might be the case with a multi-image show that has been

Figure 4.3 A properly equipped and staffed audio-visual sound studio is best way to ensure cost effective AV sound tracks. (Photo SAV Studios Ltd)

Figure 4.4 The programming of a multi-image show is the equivalent of editing in movie or video.

made for a special occasion, or a simple in-house video or slide show. More usually, show copies must be generated in the quantity and format needed to reach the audience.

In the case of slide shows, slide copies must be made from the originals, and copies of the tape cassettes prepared. Each show set should be provided with a spare tape, and show sets should each be run through on the same equipment which will be used in the field. Every item must be properly labelled so that there is no doubt how the show is intended to run. Slides should be numbered, preferably with an indelible marker pen (sticky labels come unstuck and can jam the projector) showing both the tray number and the position in the tray. If it was known at the outset that a number of show copies were going to be needed, they should have all been made when the original rostrum camerawork was done. This minimizes overall costs and maintains quality.

Film copies are made in different ways depending on the number of copies needed. The preview may have been seen using sepmag sound, i.e. the sound track was run from a separate magnetic film running in lock with the picture. For distribution a married print with optical sound will usually be more appropriate, so an optical master is made from the sound and a first 'answer print' obtained.

Final show copies are made from the negatives. The printing machine can make any necessary color corrections automatically.

It is unwise to allow the master negatives to be used too often for making show copies, because if they are damaged they cannot be replaced. When a lot of copies are required it is usual to make either a direct copy negative (CRI or color reversal internegative) or a copy negative via an intermediate positive print. The film copy printing process is then straightforward.

In comparison, the production of video copies is a simple tape-to-tape copying process. Obviously the aim is to copy from as high a quality original as possible, but it is usual not to use the actual master because, in theory, every playing will degrade it slightly. Therefore, a copy master is made and this is used whenever copies are needed. If large numbers of videocassettes are required it is best to go to one of the specialist tape-duplicating companies.

Show

On with the show. The success of an audio-visual production can depend as much on the circumstances in which it is presented, as on its innate quality. Show sponsors and producers must

Figure 4.5 If a lot of video copies are needed it is best to go to one of the specialist duplicating houses. (Photo FPA Video, London)

follow through the production to ensure that it is being used correctly. If the production is for a special occasion or fixed installation this process will happen automatically, but if it is issued in the form of a video cassette in hundreds of copies it may be more difficult to ensure that it is being presented properly.

Evaluation

Showing the production is *not* the end of the matter. Both producer and sponsor need to evaluate the response to the show.

☐ Did it meet the objectives?

☐ What was the audience reaction?

☐ If the intention was that there be a quantifiable result (fewer complaints from the public, for example) what is the quantifiable result? Is it what was expected?

☐ Was the medium chosen the right one?

☐ Did the show get the intended exposure?

Only by objectively evaluating the success of each production can users of AV gain the necessary experience to target its use more effectively.

The small-scale user of AV may regard this description of the production process as over-complicated. It is not. The same kind of procedure must be followed whether the production is a major sponsored film with a nine-month production cycle, or a simple slide/sound show made with in-house resources in a matter of days. It is only a matter of scale; part of the process that may take weeks for the big production may only take a few hours for the small, but no production steps can be eliminated just because it is a small production.

Do it yourself?

A person charged with commissioning an AV show has three choices:

☐ do all the work.

☐ get the whole show made by an outside producer.

☐ do part and subcontract the rest.

The general rule is: however simple the show, unless the organization has its own AV production department it is unwise to attempt a do-it-yourself job.

Most users get independent producers to make motivational, sales and public relations shows. Specialist training programs also need a lot of resources. However, many users do make training and product-oriented shows in-house, and many more have in-house graphics facilities for the production of visual aids. Nevertheless, when making shows in-house the experienced user will try to avoid doing everything.

A professional sound studio will complete the production of an AV sound track in a fraction of the time that an underequipped, occasionally used, in-house facility will. It can be a good idea to make a rough version of the sound track in-house to check the overall timing and effect, but the final sound track should be made in a studio. In this way advantage can be taken of the experience and facilities of the studio, including such things as a full music and sound effects library and professional narration, which not only gives a better image to the show, but saves on production time. Many sound studios catering for the needs of AV are also able to provide a programming service.

Location photography requires variable resources. If the organization has its own photographic department, it may well meet some AV needs. However AV photography requires a different approach to print photography, with respect both to format and to the number of photographs that must be taken.

The preparation of an image sequence (whether for slide/sound, multivision, filmstrip or some video) requires the use of a rostrum camera. This is especially true where successive images must register and where any graphic treatment is involved. It is generally best to get this work done by a facilities house. These studios do rostrum camera work by the hour.

Facilities houses

A facilities house is a place that will not normally do a full production, but does one aspect of the production in a highly efficient manner. The service is available on a hourly rate basis or against quotation for a particular job. Some of the services now available are:

☐ Audio-visual sound studios. Not all sound studios are suitable. Those that mainly make commercials or pop records are not always equipped to deal with the needs of AV.

☐ Slide production companies. These are at least equipped with rostrum cameras and film processing. They usually also have an art department, and an efficient means of typesetting. They produce or copy slides to order.

☐ Video editing suites.

☐ Film editing and dubbing suites.

□ Transfer suites, to convert from one medium to another.

□ Computer graphics studios. Many graphic slides can now be made on computer and then transferred to film. High-quality slides made this way need expensive equipment and experienced operators. This method of making graphics slides is gaining popularity because it is very fast and prices are coming down. Similar techniques can be used to produce video images directly.

□ Picture libraries. Conventional print-oriented picture libraries can be an expensive source. There are now AV-oriented libraries where the pricing structure is more reasonable, and the pictures are of a more useful format.

□ Film and video studios. Gone are the days when a studio was only concerned with making films for only one production company. Most studios are now rented out by the day or even by the hour, with or without technical staff.

Figure 4.6 The transfer of single screen multi-image programs to video is a well-established process. Professional studios use an optical multiplexer. This device directs the light from several slide projectors to create a single aerial image that is seen by the video camera. (Photo Vedavision Ltd)

Transfer between media

It is sometimes necessary to transfer between one AV medium and another. This may happen because a user realizes he can make a better return on his show investment by getting the wider audience that the alternative medium gives, or because a deliberate decision has been made to make a program in one medium and show it in another. For example, a training film may be shown to groups on 16 mm, but to individuals on video. Multi-image production techniques have been found to be a highly economical method of making video programs.

Some possible transfers are:

□ **Movie to video.** This is widely used. Movie film remains a highly satisfactory way of making films of the highest quality. Nonetheless, a larger audience can be achieved by also distributing on video.

□ **Video to movie.** This is not so common, and requires technically sophisticated equipment to get good results. It is used either when a trick video sequence must be integrated into a film, or when a program must be available for world distribution, where 16 mm is the one sure standard.

□ **Slides to video.** This is now very common, and is often an efficient method of original video production, as well as a means of converting slide shows intended for groups to an individual presentation format.

Slide transfers are carried out on an 'optical bench'. In principle they can be done by simple front or back projection, with the camera pointing at the screen. Back projection is better, provided a special screen with no grain is used. This type of screen material is known as CAT glass (Copy And Title).

Greater precision is claimed for the 'aerial image' method of transfer. Here the projector(s) create a virtual image which is seen by the camera. The method can be advantageous because it is easier to set up a number of projectors on the same optical axis, using beam-splitting mirrors. It also does away with any imperfections introduced by a screen.

An important point to remember when doing slide-to-video transfers is that the format is different. The original slides must be designed with this in mind. Not only does less of the slide get shown because of the format difference, but not all TV sets can be relied upon to show the whole picture. There is therefore a 'TV safe area' where it is certain that all information will be displayed. This is only about half the area of a full slide.

Low cost movie-to-video transfers are done by the back projection method, but any professional work should be done on a telecine machine. There are two types: one uses a solid-state CCD device that transfers the image line by line, and the other uses a special cathode-ray tube that scans the whole image with a flying spot. The flying spot scanner is more flexible because it gives excellent quality and also allows scanning of only parts of the image when required. This is essential when transferring a wide-screen image or zooming in on part of an image.

The transfer of video to film requires highly specialized equipment for professional results. The service is only available at a few specialist laboratories.

Copyright

Those making AV programs, even if only for in-house use, must remember that any existing photographs, drawings, designs, music and recordings are likely to be someone else's copyright. There are established methods and fees that allow the use of copyright material, but producers must *never* use existing material without having cleared copyright.

There are special music libraries for AV shows, and if these are used the procedure is very simple. There is a set scale based on the amount of music being used and on the intended audience. For example, the fee is much lower for a show that is only to be used in-house in one country, compared with one that is to be used on exhibition stands throughout the world. It is possible to use music from normal records, but this use must be negotiated with the record company concerned and is usually very expensive. If music is specially composed for a production there is no problem, but the composer may only sell the right to use the music for the one show, and further separate exploitation of the music may require extra payment.

The situation is similar with existing photographs or artwork. Unless the material is already the property of the program producer or sponsor, or is specially made for the show, a fee must usually be paid. Picture libraries have a scale fee for the use of images in AV productions. There are also some so-called copyright-free art books available; purchase of the book entitles the user to use the designs in the book for AV and graphics work without further payment.

Chapter 5

Audio-visual in business presentations

The greatest use of audio-visual outside public broadcasting and film or video entertainment is in business presentations. This chapter examines why this is, and how to take advantage of AV for this purpose. A distinguishing characteristic of this application is that although the users are professional, in the sense that they are being paid to do a job, they are *not* AV professionals or professional actors, yet often they are called upon to be both.

The *Concise Oxford Dictionary* defines a presentation as 'a formal introduction'. The essence of a presentation is *formality*, and audio-visual methods can help meet this requirement.

A possible summary of the objectives of business presentations includes the communication of *facts*, *knowledge* and *emotion* in order to:

□ **Train**, e.g. sales training, use training, job training, service training.

□ **Inform**, e.g. press and public relations activities, staff and public information.

□ **Motivate**, e.g. staff, independent representatives, customers' staff, the public.

□ **Educate**, e.g. staff, the public.

□ **Sell**, e.g. products, services, ideas.

Very few people have the ability to give a presentation just by standing up in front of an audience and giving an extempore talk. But equally there need be no-one so bad at presenting that their presentations are always failures. The trick is to understand how presentations work, and how the use of visual aids and audio-visual programs can ensure that the presentation objectives are met.

The discipline of presentations

Formality, in turn, means discipline. The idea that the use of audio-visual in a presentation means less work is, unfortunately, not true. Preparation of AV means more work at first but, correctly used, AV will ensure:

□ Better communication of the basic message.

□ That when the same presentation has to be given several times, it is always given consistently, with no accidental omissions. In this case AV is a time saver.

□ Better presentation discipline.

□ A greater likelihood that the presentation objectives will be achieved.

What is it that instinctive presenters know by intuition, and the majority find out by observation and practice? The discipline factors that affect the success of a presentation include:

□ The environment. The proper environment will greatly help the presentation, a poor environment can seriously damage it.

□ The structure of the presentation, which must be arranged to ensure that the audience retains the content. This means planning the order in which points are made or items revealed.

□ The realization that most individuals have a short attention span. The presentation must, therefore, be broken into correspondingly short sections of about fifteen minutes.

□ The realization that the pace of the presentation must be varied.

□ Rehearsal. All presentations should be rehearsed. For a major show this means a full rehearsal in the literal sense. For the small in-house event it may simply mean checking that everything is in place.

How audio-visual helps

Audio-visual programs and visual aids can make an enormous contribution to the discipline, and hence the success, of a business presentation. This applies whether the event is something as small as a training session for half a dozen people, or as

Figure 5.1 The Womens' Magazines Advertisement Presentation Department of IPC gives between 300 and 400 presentations a year to potential advertisers. All the presentations are supported by AV, and all are matched to the particular audience.

large as a new consumer product launch with an audience of hundreds or even thousands.

☐ If the environment is correct for AV, then it will also be of a kind that ensures maximum attention from the audience.

☐ The use of visual aids and AV programs demands planning. This, in turn, ensures that the correct order of presentation is worked out.

☐ AV can make a big contribution to the presentation structure by acting as a 'divider' and as a means of varying the pace of the presentation.

☐ Provided the presenter sticks to material that is visually supported, there should be no danger of straying from the point.

☐ Presentations with AV and visual aids must still be rehearsed, but the experienced presenter will find the need for full-length rehearsal reduced.

Defining objectives

Discipline relates not only to the presentation itself, but also to the purpose of the presentation. Some presentations fail, even though professionally given, because the presenter does not properly define:

☐ *Why* the presentation is being given and what the presenter is going to get out of it.

☐ *Why* the audience should want to attend and what *they* are going to get out of it.

☐ *How* the audience is expected to behave or react as a result of the presentation.

The kind of thing that often happens is that a presentation which should give a lot of facts and figures to the audience is based on a 'mood' AV show; or, conversely, an audience which needs to be motivated and excited about a new product is presented with a dreary speaker with inadequate visuals.

In the first case there could be an over-investment in the wrong kind of AV support; in the second, a failure to realize that, if the presentation objective is not going to be achieved, it might be better not to give a presentation at all.

Choice of support

Asking some questions will help the presenter determine the support needed:

☐ Based on his knowledge of the audience, how can the message be tailored to meet both the presenter's and the audience's objectives?

☐ How long should the presentation run? This will determine whether, for example, AV shows should be used to break the presentation into sections, or whether a second speaker is needed.

☐ What examples and illustrations are *available* and *relevant*?

The very act of serious advance preparation gives ideas as to the support that will help. One visual can eliminate a page of confusing script. The choice of which visual aid or AV system to use must be made in the light of:

☐ The nature and size of the audience and environment.

☐ Considerations described in Chapter 1.

☐ Available material, and available time to prepare more material.

☐ Cost.

Chapter 2 made the important point that the rules for graphic presentation are much the same for producing a flipchart as they are for making a slide. Thus, the use of AV in business presentations should be of the highest quality consistent with the occasion.

If the occasion warrants the use of a presentation room, it probably also warrants the use of slides instead of overhead transparencies. Conversely, if the presentation has to be given in an unsuitable office, an OHP may be more practical than a slide projector.

Similarly, if an expensively produced AV show is being given to an important audience, an equipment operator or technician should be on hand. If the AV show is to be shown only informally to a group of three or four people, then it is easier to transfer it to video and show it in the office.

The following chapters give more specific advice about which medium to use for the different applications. This chapter is more concerned with *attitude*. If the attitude to the presentation is right, the choice of the AV and visual aids media will not be difficult; they will almost make themselves. There is usually room for personal preference, so while the rules of good presentation must always be obeyed, there is no need for rigidity in the actual execution.

Developing script and visuals

The question of whether to develop a script first and the visuals second, or vice versa, depends on the nature of the presentation. If the whole object of the presentation is visual, e.g. the introduction of a fashion or decorative product, then it would be best to start from the visual.

However, most business presentations start from the script because the intention is to impart detailed facts. The script can then be used as a catalyst for defining needed visuals which, in turn, should reduce the script.

The organization of a presentation script is subject to the same discipline whether it is a complete script or a set of prompt notes. First define the objective, then develop a theme that meets the objective. When writing the script or script notes the presenter should observe the following order:

☐ Start by getting the audience's attention, by presenting an arresting piece of information.

☐ Tell the audience *why* they should be interested. Will what they are told save them money? Improve their job prospects? Prevent illness? Save time?

☐ State the main theme, and back it up with proof. Such proof will need visual or audio-visual support.

☐ Restate the theme, and at the same time answer the main questions expected from the particular audience. Again visual support will be required.

☐ **State the action expected from the audience.**

Some audio-visual rules

The following rules apply to the use of AV programs used as the support to business presentations. The rules apply whichever AV medium is being used:

☐ **One show, one message.** One of the main failings of business AV shows is the attempt to do too much in one show. It is more efficient to make two short shows rather than one long one. This will also help the presentation structure.

☐ **Keep it short.** Eight to ten minutes is the ideal length, fifteen minutes is really a maximum.

☐ **Use the visual element.** The second major failing is to record a written text as if there were no visuals. The visuals must be made to work. Often an original script can, and must, be cut by half.

☐ **Keep it moving.** With very few exceptions an AV show needs a new image every six seconds or so.

The presenter

However well planned the use of AV, the presenter can be a problem. There are two extreme examples of a bad business presentation:

☐ It is a big presentation. The presenter is nervous, not a good speaker, and it is almost more embarrassing for the audience to listen to him than it is for him to speak.

☐ It is a small in-house, almost routine presentation. The presenter feels he does not need to rehearse. He gives a muddled presentation, omits some items, runs out of time and leaves the audience confused, unmotivated and uninformed.

There can be some sympathy for the first, but none at all for the second. How can the presenter avoid being responsible for a bad presentation?

No one *need* be a bad presenter. Anyone who can write a good letter and who can conduct a sensible conversation *can* give a good presentation.

First presentations will always be difficult, because a lot of work is needed. As a presenter does more presentations, confidence builds up and, in many cases, parts of presentations can be reused. This, and experience, mean that the preparation work for any one presentation is reduced. For most people the rule remains: *if a presentation cannot be prepared, it may be better not to give it.*

Figure 5.2 In business presentations an audience of six can be at least as important as an audience of 600. (Photo courtesy Missenden Abbey Management Training Centre)

Presenters can be nervous about speaking in public; they should be worried if they are *not* nervous. All the best actors and speakers suffer from butterflies before they go on, but, equally, they know what it is that they are going to do and say. They are *prepared*.

The audience

There is no presentation without an audience. In business an audience of six people may be as, or even more, important than an audience of 600. Successful presenters always put themselves in the position of their audience when planning their presentation. The attitude 'What's in it for me?' may sound selfish, but if the audience are listening to something that they want to hear they will happily tolerate the less than perfect speech.

An audience always starts by being on the side of the presenter, they *want* him to succeed. The presenter can benefit from this, but not take the goodwill for granted.

Script or notes?

The best presentations are those where the speaker seems to be giving an impromptu speech, and yet covers the ground and finishes exactly on time. These are the ones that take the most time to prepare!

If a presentation is being given for the first time a suggested procedure is first to write a complete script. This will ensure that everything is in the right order and eliminates unnecessary repetition. Then, when the presenter is completely familiar with the script, he should make a set of key notes and give the presentation from these notes. The important thing when actually giving the presentation is to stick exactly to the original script concept. If a carefully planned sequence of visual aids is used the aids themselves can act as the notes.

If a section of a presentation demands tight scripting and strong visual support, this part should be given as an AV show.

Why rehearse?

The single most important element of success in presentations is *rehearsal*. There is nothing worse for a presenter than finding himself in front of a visual he cannot recognize, or running out of time.

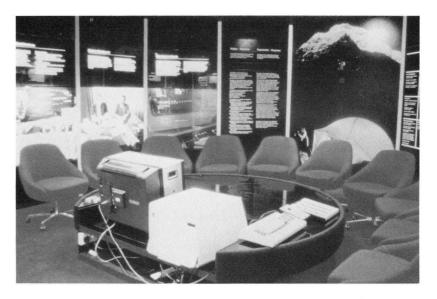

Figure 5.3 The Scottish Development Agency (SDA) gives specialized presentations to potential investors at its Glasgow Headquarters. The presenters sit with the audience. (Photo Mackay Design Associates)

The first-timer giving an important presentation should have a private rehearsal for self-evaluation. He should then ask one or two colleagues to sit in on a second rehearsal and invite their comments and suggestions.

'On site' the presenter must be completely familiar with his visual aids and AV shows. He should insist on having a technical rehearsal, and always remember that the audience are collectively spending a lot of time to attend the presentation. As a proportion of their man-hours devoted to him, his rehearsal time is small.

Table 5.1 Dos and Don'ts for presenting yourself

Do	Do not
Keep it simple	Fidget or wave your arms
Speak clearly	Mumble or drop your voice
Look at the audience	Look at the ceiling
Keep to time	Over-run
Keep to plan	Ramble
Prepare properly	Apologize
	Talk to the screen
	Wave the pointer around
	Jingle your loose change

Figure 5.4 The SDA use slides, movie, video display panels and direct computer output to support their presentations. (Photo Mackay Design Associations)

Chapter 6

Audio-visual in conferences

In the context of this book, conferences include two types of event. First, the general conference of a professional association, or one set up by a commercial organizer on a specialist subject using invited speakers. Second, the sales conference or product launch, where a company or organization calls together a group, usually of employees, sales representatives, dealers or customers, who are in some way financially interested in the subject matter.

It is no surprise that it is the second type of conference that is most suited to AV methods. Company sales conference organizers have come to realize that these conferences benefit from AV because:

☐ It helps provide a sense of occasion.

☐ It helps provide a structure to the meeting.

☐ It is cost effective, because it can save time and impart correct information efficiently.

☐ It can be a good motivator of people.

The use of AV methods is not a new idea. The USA led the way, because early on they had a unique problem. Any company trying to operate nationally was forced to bring the message to many outposts who might be not only many miles, but many days away. The sales meeting was developed as a means of motivating the distant markets, and lantern slides and show business were pressed into service. A set of lantern slides from the early 1930s, used by the Pacific Gas and Electric Company to get the salesmen moving, can still be seen today and their quality would be immediately recognized by any modern conference producer.

The Screen Works in Chicago, now a well-known manufacturer of collapsible fold-up screens ideal for conference work, was originally Wilcox Lange Inc. The splendid partners, Herb Lange and Kay Kallman, started the company in the late 1950s after nearly a lifetime doing road shows for the automobile companies and other consumer product companies. They did not have the benefit

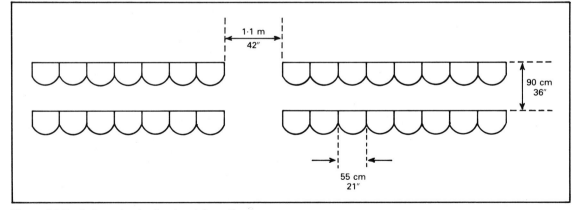

Figure 6.1 Typical spacing for theater-style seating in an auditorium or conference suite.

of the automatic Carousel™ projector, so had to work like dervishes to feed hundreds of slides manually through 'push-pull' projectors, while operating hand-operated mechanical iris dissolve mechanisms. Ironically, they were able to project brighter images then are seen nowadays, using highly efficient optical systems with $f/1.5$ lenses.

The fact is that sales conferences are, and always have been, a branch of show business. Organizers should recognize this and plan the whole event accordingly, but should not lose sight of the objective(s) of a particular event. The well-known theater critic's line, spoken on emerging from a well-staged but tuneless musical, that he 'came out whistling the scenery', can apply to sales conferences that have gone over the top. As in all things, organizers should be less concerned about using the latest gimmick – be it laser, high-speed computer graphics, videowall or multi-image show – than with what they want the audience to do as a result of having attended the event. If this is clear, the choice of medium will be almost automatic.

Choice of media

The problem with sales conferences is that they are one-of-a-kind events, and often the information to be given is not available until a few days before. In principle, all the AV media and visual aids systems can be used, but in practice slides are the main basis of conference AV, for the following reasons.

First and foremost, slides can give very high quality visuals that can be seen by large audiences. A single slide image can comfortably be projected using simple equipment costing a few hundred dollars. To get the same image brightness from, for example, a graphics computer via a video projector needs equipment costing over $100,000. As a compromise, equipment giving half the Carousel™ image brightness still costs in the region of $20,000. The image quality cannot be as good in either case.

Chapter 20 describes systems that allow on-line sequential display of computer graphics. These systems do have a place in a permanent presentation room, but there has recently been a tendency to introduce them into sales conferences. There *may* be an argument for doing so – for example, if the designer has elected to use a very high tech set with multiple video screens. However, if such systems are proposed, the conference organizer should fully understand what the system is going to do that a slide-based system cannot. There have

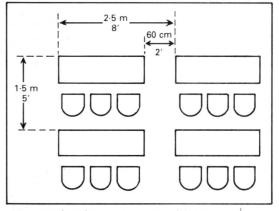

Figure 6.2 If conference or seminar delegates must be seated at tables, this is a typical arrangement.

been a number of cases where ludicrously expensive equipment has been used on site (including complicated video multiplexers to get a slide image that could well have been projected directly onto a video screen): it would seem that the poor customer has been paying for the producer's over-investment in unnecessary equipment.

Most conference work consists of speaker support, which requires good quality *still* images. So, it might be argued, why not use the overhead projector? It may possibly be suitable for the less formal sessions, but it is an untidy method for the main event, and is unable to give the smoothness of slides. The very discipline of having a slide sequence designed and made contributes to the overall polish and success of the event.

Slides are the most flexible medium because:

☐ The same system can be used for speaker support, visual aids and automatic audio-visual programs.

☐ The same system can show text, graphics and high-quality photographs.

☐ The projection system can be easily adapted to match the environment. Big screen images can be achieved at low cost using multiple projectors.

☐ The wide choice of lenses available make slide projection easy to design to match a set or a venue.

☐ Changes and additions are easily made.

The needs of the smaller sales conference are well met by a two or three projector single-screen system. This would be manually operated for the speaker support sequences, and run from tape when showing pre-recorded AV programs or modules.

The larger conference or product launch will usually use multi-image with nine, twelve, fifteen

Figure 6.3 Major sales presentations are often referred to as 'Industrial Theater'. (Photo Caribiner Inc.)

or even more projectors. This is partly to give greater impact to the pre-recorded sections, but mainly to allow the use of a screen area big enough to match the audience. For example, many hotel banqueting suites have relatively low ceilings, which may allow only a 2 m (6 ft) high image. A single 3 m × 2 m (10 ft × 6 ft) screen is inadequate for a large audience. The wide-screen look of, say, a 6 m × 2 m (20 ft × 6 ft) screen is more appropriate to the room, and allows new possibilities, such as comparisons and panoramas, on the screen.

Modern computer-controlled multi-image systems are very reliable, and allow mixed live and recorded sections. Speaker support can be as animated or complex as required. They also allow for those last-minute changes which, in a perfect world, would not be needed, but in practice are. Their operation is described in more detail in Chapter 16.

Very often an existing video program must be run as part of a conference schedule. If the program exists only as a video program, there may

be no choice but to put in the largest video projector that can be afforded and run the show from video tape. However if, as is often the case, the show was originally made on movie film or slides, it will have much more impact for a large audience if it is shown in the original medium. The definition of a large audience will vary according to its importance, but if a screen size bigger than 2.5 m (8 ft) wide is required to give an adequate image size, the optical alternative should definitely be considered.

Industrial theater

Modern sales conferences have been referred to as 'industrial theater', and the largest of them may well use greater resources than a Broadway musical. They can be great occasions, so it is vital that they are well planned and executed. Because the cost of the event itself is often small compared with the cost of bringing the audience together (not to mention the cost of their time) it is usually best to call in an outside conference producer to coordinate the event.

Some companies are big enough to have departments whose sole job it is to organize conferences, and who may even have their own staging facilities for the smaller conference. Even they, however, would not attempt the big event without outside assistance. There is a range of firms to choose from. At one end there are theater-oriented companies who have the resources to mount a multi-million dollar launch of a new car. At the other there are specialists who understand the needs of small industrial sales conferences, and who will organize all aspects of staging, preparing visuals, making AV program modules, and so on.

The sense of occasion is important. The smallest conference should use a properly dressed screen, discreetly placed equipment with no trailing wires, correct lighting and sound – all items that can only be ensured if the venue has been thoroughly researched beforehand. Most conferences benefit from the use of a conference set.

Simple sets consisting of a screen, side flats and a matching lectern can be hired, and are easily customized to include the user's company logo or conference theme. The more elaborate occasion demands the use of specially designed sets, and it is here that the talents of theater people are required.

Figure 6.5 Sometimes the set is designed to reflect the theme of the conference, as in this classical set for a Legal and General Assurance Co. convention in Athens. (Photo Roundel Productions Ltd)

The set should be seen as the item that brings cohesion to the whole event. However simple it is, as soon as they see it, the audience know that the presenters mean business. It is all part of getting the most out of the presenters, the audience and the event itself.

Some conference rules

Several suggestions for the smooth running of conferences can be made. The same rules apply, however big or small the event.

The first rule is to avoid a situation where one person speaks for more than 15–20 minutes at a stretch. If one person must speak for a long time, the presentation should be broken up, either by alternating with another speaker or speakers, or by the introduction of AV modules. Such modules need only be a few minutes long.

This highlights a common problem. Often the in-company conference organizer, or even the outside producer, is afraid to tell the presenter what to do, because the presenter is the Chairman of the Board or some other exalted personage. They simply let the poor fellow go ahead and give his rambling 90-minute speech because he asked for, or took, a 90-minute slot. Presenters should listen to outside advice given by people whose business it is to script and run sales conferences, because they have the experience to know what motivates audiences, and how to get the right mix of events within a presentation. Very often what the presenter perceives as important (and which may take up 70 per cent of his proposed slot) may not actually be helping the conference objective.

Figure 6.4 A conference set makes a good impression. Here a hired set has been customized. (Photo Presentation Systems Ltd)

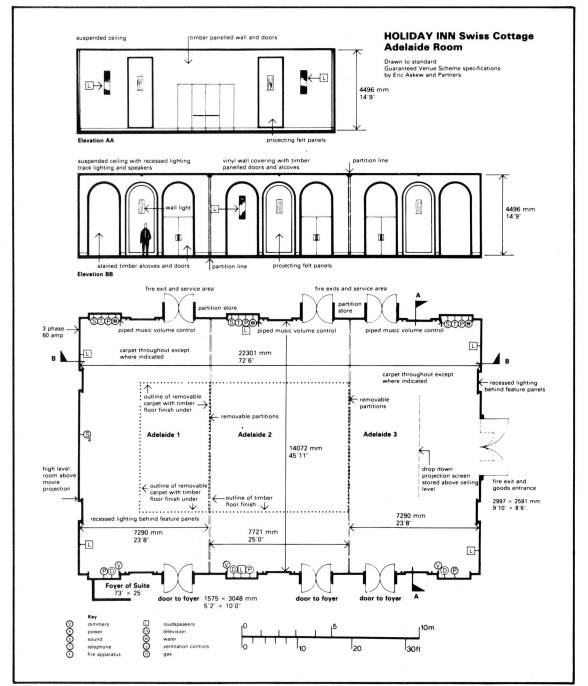

Figure 6.6 The conference venue should be able to provide an architect's drawing, so the organizer can ensure that everything and everyone fits in. Drawing reproduced from *The Conference Blue Book.* (By permission of Spectrum Publishing and the Holiday Inn, Swiss Cottage)

Introducing too much detail is a sure recipe for failure, because the audience cannot absorb it and will quickly lose interest. This type of information should be given in a different way, perhaps as part of the written support material. However, it may be better discussed in the more appropriate environment of a parallel seminar.

A speaker should always have visual support but there will sometimes be sections of his presentation where visual aids are redundant. In these cases it is important not to leave up an irrelevant slide, but equally important not to have a blank screen. The company logo, conference theme slide or presentation title can be used to fill in.

Some organizers insist that their presenters never have a slide on the screen for more than 20 seconds. This is impractical for most users, but does make the point that the amount of information on any one visual should be limited.

All conference events should be properly staged. This means that even in the simplest event, using just one slide projector, there must be someone whose job it is to look after the AV. With the possible exception of single-slide advance, the presenter should not operate *any* equipment.

The conference venue should be thoroughly researched. The conference producer will do this on behalf of a client, but if a venue has to be selected before a producer is appointed, the following minimum requirements must be met:

☐ There must be proper blackout.

☐ The lighting must be under dimmer control. This should be the push-button automatic type which allows control from different places within the room.

☐ There must be easy access.

☐ The room must be big enough to accommodate the intended audience, and allow them to be within the optimum viewing area for AV. (See Chapter 11 for these standards.)

☐ There must be sufficient height to get in a big enough screen. If necessary it must be possible to move chandeliers or other obstructions to projection beams or sight lines.

☐ Unless the conference company is providing a separate system, there must be a good speech reinforcement (PA) system. Microphones and cables must be in good condition, and back-up service available on site.

☐ If there will be no special conference set there must be a good lectern, with separate lighting to light the presenter and not the screen.

A good rule to observe is that if the intended conference center or hotel is unable to provide an architect's drawing of the room to be used, showing all power points, controls, obstructions,

access etc., it can be safely assumed that their venue should *not* be used.

Finally, time must be allowed to have proper rehearsals. The sales conference is, or should be, a highly structured event. Only by full rehearsal can everyone participating be sure that the event will work as a whole, in respect of content, timing, logistics and staging.

AV and the general conference

Unfortunately the general conference rarely has the organization to permit more than the simplest visual aids (slides or overhead projector) to be used, although many would benefit from better integration of AV methods into their proceedings.

Speakers at these conferences generally find that they have to fend for themselves. They are given a time slot to fill, and are expected to perform at the stated time without rehearsal. They must follow the rules on presenting themselves outlined in Chapter 5, in particular:

☐ They must have their own private rehearsal to ensure that their presentation fits the time slot.

☐ They should make maximum use of slides as visual support, taking special care that the slides are legible at long viewing distances.

There are a number of commercial conference organizers who run special-interest conferences where they get together expert speakers. Some of these organizers are exemplary in ensuring that the event makes the best use of the audience's and presenter's time. Others could be accused of negligence, if not daylight robbery. They should run such conferences to the same discipline as outlined for the company sales conference, but usually do not.

A conference that is set to take place over several days may well be a means of justifying a bigger fee. In fact, most busy people would prefer a shorter, more effective conference – even if the fee was still the same.

Organizers of this type of conference *should* (but often do not) ensure that their speakers are chosen for their known ability to present well to an audience. The use of visual aids, prepared to a uniform high standard, should be mandatory, and the organizers should work with the individual presenters to assist with their production or else should employ a conference AV producer to help.

The whole point of these conferences is that participants should gain more by attending than they could by simply getting a copy of the written papers. There is no point in a presenter reading

Figure 6.7 Stage managing a major conference is as complex a task as staging a musical. (Photo Spectrum Communications Ltd)

from a written paper that is then handed out to the audience. The written paper should give all the detailed support, and should be written to be read. The presented paper is part of a perform-ance, and should be written to be performed. The performance cannot be just the spoken word, it needs visual support. The aim should be to give the audience an overview of the information needed to allow participants to decide how they should proceed.

Prompting systems

Many stage-managed and fully scripted confer-ences use television-style automatic prompters such as Autocue™. These allow the speaker at a lectern to appear to know his scripted speech off by heart. The audience cannot see that he is, in fact, reading it from a virtual image seen as a reflection in one or more clear glass plates which

only he can see. It is best to rent these systems, because to be effective they need a skilled operator.

Traditionally, automatic prompters have used the script prepared on a long roll of paper with large lettering. The roll is scanned by a TV camera, and the operator ensures that the motor-driven roll is kept in step with the presenter. More recently, these systems have become computer-ized. This has the great advantage that script sections can be prepared and edited as on a normal word processor. A special computer graphics program then ensures that the text is scrolled and displayed in a large format with a font legible at a distance.

Whether these devices are desirable throughout a conference is debateable. Many of the best presenters work best from notes, and spontaneity can be lost by automatic prompting. However, in a major product launch using complex staging they may be essential, if a whole stage crew is depending on the presenter to read the script accurately and consistently.

Incidentally, the same secret viewing technique can be used for giving the presenter a view of what is on the screen behind him. One of the problems that conference producers have is stopping presenters continually turning round to see what is on the screen; they need reassurance that the right visual is there! Incurable cases can be helped by the head-up video reassurer.

Figure 6.8 An automatic prompting system in action. The speaker can see his script in the plain glass screens in front of him. The image is invisible to his audience. (Photo for Autocue Ltd by Srdja Djukanovic)

Chapter 7

Audio-visual in training

Training in business and commerce, further education and deployment of labor has one or more of the following objectives:

☐ The briefing of an audience or individual on a situation.

☐ The teaching of a specific skill.

☐ The imparting of specialist knowledge.

☐ The imparting of knowledge about a company's product.

☐ The imparting of knowledge about an organization's services.

☐ Training in procedures.

☐ The teaching of preferred behavior.

Audio-visual methods can make a contribution to all of these. This chapter covers some topics relating to AV in training, although because training is one of the main uses of AV, most of the other chapters are also relevant to the subject.

Training is a structured process, so AV is a help in providing that structure. However, there is merit in simplicity, so trainers should use the simplest AV method that meets their objective, modified by considerations of distribution of the training material.

A specialist engineering company might consider the use of short video programs to train their staff how their products work. On further examination of the training problem, they might conclude that their needs would actually be better met by the use of the slide/sound technique. Apart from the cost savings, the alternative method would allow for easy updating and program modification, and would realistically allow short programs to be made in-house.

On the other hand, a retail store group might have a simple training program on cash handling procedures that was easily made on slide/sound. However, they then have the problem of delivering the program where it is needed, and conclude,

rightly, that the simplest device to have installed at every branch is a TV set with video recorder. Thus they might create the training sequence on slide but distribute it on video.

Most training needs are highly specific. This means that there is rarely a big budget for the preparation of training materials, although there should be a *regular* budget which, if wisely used, will continually develop the training resource.

The training room

The success of audio-visual depends as much on a commitment to its proper use, as to making the appropriate kind of program material. Anyone intending to use AV methods for training should ensure that a suitable space is available for them to be effective.

For some organizations the training room will be the same as the presentation room which is described in detail in Chapter 11. Many users will need only a training room, some will need both. In principle, the training room is a simpler concept than the presentation room, and its prime requirements are:

☐ that it is big enough for the expected training groups;

☐ that basic visual aids such as writing boards and screens are permanently installed;

☐ that there is proper provision for the expected AV equipment so that it does not distract. For example, there should be a separate projection room, or at least a projector stand at one end of the room equipped with the *right lenses* to avoid equipment in the middle of the room;

☐ that there is simple lighting control to ensure that projected material can be seen properly, and that notes can be taken.

In other respects the training room is less formal than the presentation room. The important thing is that the use of AV material should be easy and require no setting up.

Figure 7.1 The training room at the Lutheran Brotherhood Insurance Company, Minneapolis.

The overhead projector

An almost essential element of the training room is the overhead projector (OHP). Correctly used this simple device makes a major contribution to training presentations by providing easy-to-use visual aids.

The preparation of material for the OHP can be done in many different ways, most of which are inexpensive and quick. The OHP can be used in a fully lighted room. If there is no training room or if a training session has to be given in a less than ideal environment, the OHP is probably the best visual aid system for overcoming environmental shortcomings.

The OHP is a very personal method of presentation support, because the presenter can face his audience while operating the machine, and can do so either sitting down or standing up. He can point to an item on the transparency with a pencil without having to turn round to the screen. A correct sequence of OHP transparencies gives structure to his presentation, and lecture notes can be carried on the transparency frames to give further fluency to the presentation.

The OHP can be used as a writing board because the presenter can write directly on to a transparent foil with marker pens. Some people use two OHPs at a time, because this allows them to use one to display formal graphics, and the other as a writing board.

Animation kits are available to produce simple animation on the OHP transparency. For example, they can demonstrate mechanical movements and fluid flow. Items like mechanical linkages can be demonstrated directly by simple two-dimensional moving graphics. Fluid flow is best demonstrated using the polarizing technique. Here the light beam is interrupted by a rotating polarizing disk, which continually changes the plane of polarization of the projected light. The OHP transparency itself is conventional, except where the flow is to be indicated. A layer of special film is superimposed over these places.

The special film consists of strips of polarizing material with alternating or varying planes of polarization. The effect of viewing this via the rotating disk is that the treated areas appear as travelling dark bands. The direction of motion depends on which way the material has been laid down.

The skilled OHP presenter quickly learns tricks that can be used to improve the presentation style. 'Successive reveal' can be done simply by covering the transparency with an opaque card, and sliding it away as each new point is to be revealed. 'Build-ups' are achieved by starting with a clear transparency that only carries the basic informa-

Figure 7.2 The overhead projector combines the facilities of a writing board and a projector. (Photo Bell and Howell AV Ltd)

tion, for example, the coordinates of a bar chart. Successive transparent overlays are then added, each entirely transparent except for its add-on information, such as one element of a bar chart.

The principle of the overhead projector is shown in Figure 7.3. Most projectors use a tungsten halogen lamp, although very expensive compact arc-lamp models are available for those who want big images, up to 8 m (26 ft) square, in large lecture theaters or briefing rooms. Behind the lamp is a concave mirror, and in the optical path ahead of it is a large fresnel lens. A fresnel lens is a light-gathering device made of glass or plastic that is molded with a carefully calculated linear or concentric prismatic pattern. In the OHP it serves

as a condenser lens. A conventional condenser lens of the size required would be very heavy and expensive.

The transparency is placed on a transparent platform just ahead of the fresnel lens. Next in the optical path is a mirror that is used to divert the light beam, and finally there is an objective projection lens, usually with a focal length around 300 mm. In the best models the mirror is in a sealed unit. Some models reverse the order of mirror and lens, to make a device that folds up into a smaller space, but care must be taken to avoid damage to the mirror, and to keep the lens clean.

Originally the stage on an OHP was 250 mm × 250 mm (10 in × 10 in), but in recent years projectors have moved over to A4 format (210 × 297 mm) which requires a stage of 297 mm (11.5 in) square if both landscape and portrait format transparencies are to be used. Transparencies in 250 mm × 250 mm (10 in × 10 in) card mounts will project on either type of machine, but A4 transparencies will be cut off on the long picture side when shown on one of the older machines.

The overhead projector is a relatively simple device, but there is a wide range of machines to choose from with a correspondingly wide range in price. It is not worth paying more than is necessary to do a particular job, so the benefits of special features offered should be evaluated accordingly:

☐ The most economical units use lamps like the 24 V 250 W one used in the Kodak SAV projector. This is an obvious advantage, but if very big images are needed (more than 2.5 m or 8 ft square) larger lamps are necessary.

☐ The more expensive projectors have more even illumination and better resolution. Projectors should definitely be evaluated for evenness of illumination, but high resolution is not needed if most of the material to be shown is bold graphics.

☐ There is a choice of lenses within a narrow range, typically 280–360 mm, but special lenses of extra long focal length are available.

☐ Features such as twin lamps with a fast changeover switch, lamp economy switch, anti-glare filters and ease of stowage should be evaluated only in so far as they are of benefit to the particular user.

One important aspect of the use of the OHP is that, because the projector is normally sited so that its lens is below the bottom edge of the screen, the projection beam may not be at 90 degrees to the screen surface. This will introduce keystone distortion to the projected image. The problem is eliminated by tilting the screen forwards as shown in Figure 7.4.

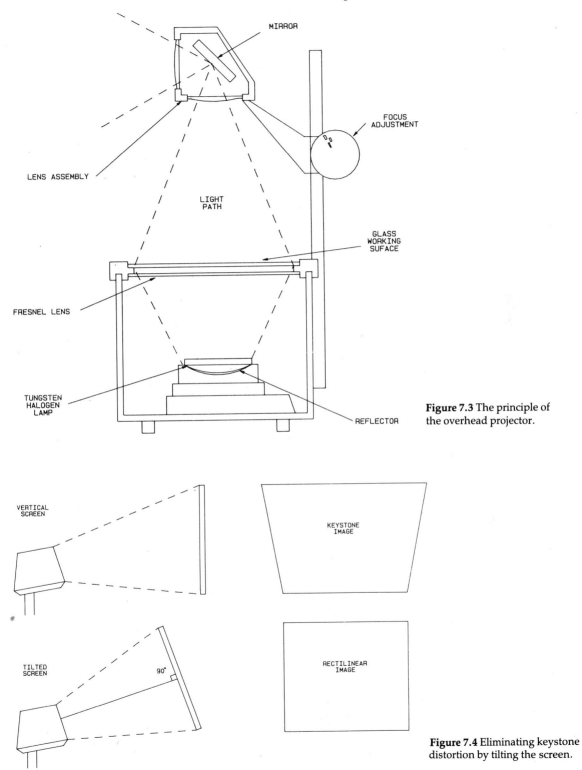

Figure 7.3 The principle of the overhead projector.

Figure 7.4 Eliminating keystone distortion by tilting the screen.

Figure 7.5 A teaching wall that combines various writing and projection surfaces makes a training room easier to use. Notice in the foreground furniture specially designed for the training room environment. (Photo Teaching Wall Systems Ltd)

Figure 7.6 A video trolley provides both security and convenience. (Photo Elf Audio-Visual Ltd)

Training room furniture

There are now a number of furniture items that are simple in concept and can make a useful contribution to the success of a training room. Their usefulness will depend on the user's exact training aims and training methods.

The 'teaching wall' is available either as a custom-made piece of furniture, or as a standard item. These walls contain writing surfaces and different types of screen. If back projection is used they may also house the equipment.

The 'video trolley' is a secure trolley which houses a video recorder with accessories and a large monitor. It has large castors to allow easy movement between different rooms.

Many of the top office furniture suppliers have chairs and tables intended for the training room. The tables can be arranged either as individual desks, classroom style, or linked together for meetings and discussions.

Where a full teaching wall is not justified, the installation of a 'picture rail' probably will be. There are rail systems available that allow the easy hanging of screens, flipcharts, writing boards, display panels and even equipment. The items are supported from the rail, and they can be easily slid to different positions as required.

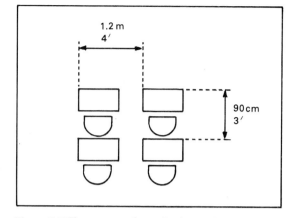

Figure 7.7 Classroom-style seating in a training room cannot be closer than this.

Figure 7.8 The combined meeting and training room sometimes uses a rail system for AV aids (as opposed to a fixed teaching wall). This is the Bretford Walltrak™ System.

The study carrel

Sometimes when self-instruction is a part of training, the 'study carrel' (a desk at which one person works and which has one or more means of AV communication built in) may be of use.

These systems are highly effective, but *only* when there is the proper software backup and commitment to the continuing development of programs. In practice, this means that either the organization is large enough to have a department devoted full time to the production of program material, or it is necessary to develop a strong connection with an independent specialist producer of this software.

The success of these kind of programs depends on the continuity of the relationship between the organization which needs the training and those who create the programs. The nature of training programs is quite different from those made for sales promotion!

There is a choice of medium, and a real danger of choosing a medium for its apparent virtuosity and technical sophistication, when a simpler method would suffice. The standard medium for the study carrel has been slide/sound, and for many users it will remain so for a long time yet. This is because it is the most practical medium for making and updating programs in-house, especially where the number of trainees who will see any one program is small. The equipment need only consist of a simple AV cassette recorder, a slide projector and (usually) a rear projection screen arrangement. It can even be one of the self-contained units used for individual presentation.

As soon as the number of trainees increases there may be an argument for moving to videotape. At its simplest, this means transferring what would have been the slide program onto video, with, of course, the added possibility that movement can now be included.

Most program makers are best at making linear programs which run from the beginning through the middle and to the end. For many users it can be easier, and a better investment, to make a lot of short linear programs which each make a single point, than it is to make complex programs which include the possibility of branching, multiple choice and student assessment.

For the brave, however, all these things are possible at quite reasonable cost in equipment terms. The subject of interactive audio-visual programs is sufficiently important to warrant a chapter of its own, so further discussion is reserved for Chapter 12.

Figure 7.9 A typical study carrel. (Photo Synsor Corporation)

The training film

Most company training is specific to the company, but training in behaviour can be universal to many companies. For example, training in aspects of safety, hygiene, selling, general management, finance, service, customer relations and security could be the same for a company making electronic equipment as for one making tents. Many excellent training films are now available. They can be rented or purchased, and are usually available in both 16 mm and videocassette.

If these films are rented for a short duration they must *never* be copied. They are copyright material, and apart from the immorality of copying, the penalties are high. Not only is there a heavy fine, but the film companies make a point of advertising the names of offenders. If a particular film is used frequently, it should be bought.

Films should not be used in isolation. They should be used to provide structure and to vary the pace of an overall presentation on the subject. Often the suppliers of the films provide written support material to help the user make the best use of them.

If there is a large audience it is important to show the film in the right environment with the correct lighting conditions. An audience of 20 or more will be much more impressed by a big bright 16 mm movie image than they will be by all trying to look at a 22-inch TV set. As a group seeing the film together, they will also respond to and absorb the training message better.

Figure 7.10 Training films are availble on a wide range of subjects, from telephone selling to general management. (Photo Video Arts Ltd)

AV programs

Besides readily available material such as training films, specially made AV programs for group showing (as opposed to visual aids and programs made for individual training) may be needed. These programs should be short, and deal with only one topic at a time. They should be designed as an integral part of a training session, and be of a form that is easily updated. A point that needs repeating is that the production of many short programs is usually more effective than producing a few 'Ben Hur' length programs.

Figure 7.11 The National Motor Museum at Beaulieu in England uses a slide/sound program to help with the initial training of staff. Because they take on large numbers of temporary staff in the summer, they need a system that is easy to use and easy to update.

Role playing

Another aspect of behavior training is to get trainees to act out a part, for example a salesman, stewardess or bank clerk. This type of training is helped enormously if the trainees can see themselves. This can be achieved by simple video systems that, in principle, need consist of no more than a low-cost video camera and a video recorder. It is more important that the circumstances of the recording match those of the intended area of use (office, shop etc.) than that the lighting and camera facilities are of a high order.

It is important to understand that these simple video systems are *not* suitable for making actual video programs. There may well be some users who can justify a full video production set-up, including a small studio, for in-house use. Of course they can then use this for the role-playing training, but care should be taken to ensure that the trainees are not intimidated by the artificial environment of the studio, and the training procedures should stay firmly in the hands of the training director and not get into the control of the AV producer!

Networks

Large organizations such as banks and retail chains have the problem of training large numbers of staff each year. Some of the higher-level and specialist training should still be done at training centers because attendance at the center, and meeting with participants from other parts of the organization, will increase the effectiveness of the training process. However, most induction training, low-level training, and training in new procedures and products is best done at the place of work. In the case of a bank this can amount to several thousand sites.

Such large-scale trainers have to take several things in to account before deciding on the precise method of dealing with the problem. The logical solution is to equip each branch with a video monitor and to provide program material either in the form of videotape or videodisk. The main problem is to ensure that each site has a suitable room for the training system and that the local management is committed to using the training materials provided – in particular, that they allow the staff the correct time to use the system.

The question of tape or disk is an open one. If a large number of branches are involved, disk could be both more economical and easier to use because random access of program segments is extremely fast. It is also much better for those applications which need still frames. Another consideration is the possibility that the video system used for staff training is also used by the public as part of a point-of-sale display. This idea is explored further in Chapter 12.

Modern video equipment is reliable, but may need service like anything else. Those contemplating video networks should ensure that local suppliers have an interest in the sale to ensure after-sales back-up. This means either using multiple suppliers, or one main contractor who undertakes to organize local service on a subcontract basis.

In spite of the advantages of video as a multi-branch training medium there are other options, especially for the smaller user. Portable slide/sound units or standard projectors with an AV cassette recorder may well meet the needs of the smaller chain better than video, especially where it may be necessary to modify programs locally.

Student response

Some organizations must train groups of people on courses. Examples include the airline and catering industries. In these cases it is useful for the instructor to know how much has been understood by the group.

Student response systems can be used by the instructor both to modify his own presentation in the light of the response received; and as a means of individual student assessment. These systems vary in complexity, but all essentially consist of an automated multiple-choice question system.

Each student has a set of push-buttons, typically 3 or 5, corresponding to the maximum number of choices likely to be required. When the instructor wishes to assess the group he presents a question,

Figure 7.12 A student response system is useful in some kinds of group training. These are typical units, one for each student linked by a single cable to a central computer. (Photo Reactive Systems Inc.)

Figure 7.13 The flight simulator represents the highest level of audio-visually assisted training. This is the outside of a Boeing 767 Simulator. (Photo Rediffusion Simulation Ltd)

Figure 7.14 A flight simulator from the inside. Notice the wide-screen computer-generated display of the runway. (Photo Rediffusion Simulation Ltd)

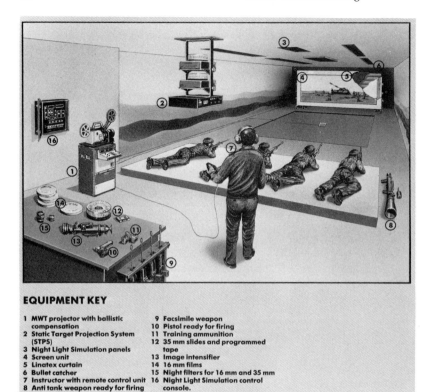

EQUIPMENT KEY

1 **MWT projector with ballistic** **compensation**	9 **Facsimile weapon**
2 **Static Target Projection System** **(STP5)**	10 **Pistol ready for firing**
	11 **Training ammunition**
3 **Night Light Simulation panels**	12 **35 mm slides and programmed** **tape**
4 **Screen unit**	13 **Image intensifier**
5 **Linatex curtain**	14 **16 mm films**
6 **Bullet catcher**	15 **Night filters for 16 mm and 35 mm**
7 **Instructor with remote control unit**	16 **Night Light Simulation control**
8 **Anti tank weapon ready for firing**	**console.**

Figure 7.15 Audio-visual methods are used in firearms training for the military and police. (Diagram GQ Defence Equipment Ltd)

usually by slide or video projection, together with the numbered answer options. The students respond by pressing the appropriate button.

Depending on the type of system installed the instructor can see the totals for each answer, and/or can see individual responses. A record can be kept of each individual's response if necessary. The possibilities are extensive, but again the system is a waste of money unless there is a commitment to using it properly, and to generating the course material that makes it possible to use it at all. In those countries where there is a written, as well as a practical, driving test this style of audio-visual group examination may eventually replace the written paper, especially because it allows motoring situations to be displayed to the examinees along with relevant questions.

Some advanced training facilities now have a personal computer at each trainee's position. At present this is usually only the case for training related to computers and computer methods, but it is likely that some trainers will consider this a valid approach for other types of training. The communication then becomes two-way. The in-

structor can download programs and course material to all students, but can also elicit response or examine individual progress as required.

Simulators

The highest level of audio-visual training is probably encountered in the simulator. The best-known type of simulator is the flight simulator, where the most sophisticated versions may cost many millions of dollars. The flight simulator consists of a cockpit exactly like that used on the aircraft being simulated. It is equipped with a video projection system that presents the view out of the cockpit window, and the whole assembly is mounted on a set of hydraulic rams that can simulate take-off and landing accelerations, runway rumble, stalling etc.

The whole system is designed to fly like the real aircraft, and is used by pilots both for initial training and for re-training. Full-length flights are flown, and the simulation is so realistic that participants often feel that they really *have* landed

Figure 7.16 The static target projection system used in firearms training. (Photo GQ Defence Equipment Ltd)

in Toronto or wherever at the end of their 'flight'. Apart from routine training the simulator allows pilots to react to simulated situations that would be too dangerous to try on a real aircraft.

These types of simulator bring together the latest in high-speed computer graphics (for constructing the moving video image), video projection (for presenting it), and computer control and mechanical engineering. They are used for all types of civil and military aircraft and spacecraft.

The same idea can also be applied to simpler applications. Ship simulators are used for training tanker and freighter crews and for training ships' pilots, especially when a very crowded harbor or difficult approach is involved. It is likely that simple car simulators will soon be widely available, both for fun and as a serious way of introducing people to driving.

Relatively simple simulation methods are used in fields such as firearms training. It is obviously better for live ammunition training to be done on a simulation basis!

Chapter 8

Audio-visual in selling and public relations

There are only two reasons for using audio-visual techniques in selling:

☐ Reducing the time taken to make a sale.

☐ Increasing the probability that a sale will be made.

If the AV show is not able to meet either, or both, of these objectives, then it is the wrong show for selling, even if it is the right one for staff training.

Person-to-person sales presentations

Audio-visual programs are being used successfully in individual sales presentations in many diverse applications. In all cases it should be possible to achieve *both* major objectives, and it must be remembered that it is often just as important to reduce the overall selling time from the buyer's point of view as it is from the seller's.

AV programs for person-to-person selling should be designed to reinforce what the sales person has to say and to give information in a way that he or she cannot. They should also give a structure to the salesperson's presentation. Therefore, it is no use making such programs unless the exact circumstances in which the show is to be given are taken into account. It is very important *not* to take over the salesperson's job, and the show must not attempt those elements of selling, such as 'closing', which can only be properly done by the salesperson.

The examples below illustrate those kinds of selling situations that can be helped by AV methods.

A manufacturer of drinks-dispensing machines wanted to enlarge his share of the hospital market. His competition was not the other manufacturers of vending systems, but the existing method of drinks dispensing in the hospitals. This was based on using traditional methods, requiring crockery

which had to be washed, and therefore the service of hot drinks was limited to set times of day.

The manufacturer realized that he had to appeal to the operators of the traditional system. He had to overcome prejudice against machine-made coffee and tea, and the idea that dispensing systems using disposable cups were more expensive per cup than the traditional system. He therefore financed a trial at one hospital to allow the hospital management to get direct experience of the alternative. The trial was successful and a sale resulted.

It would have been uneconomic and a waste of time to repeat the process for all potential customers. So a video program was made which compared the two methods, and, most importantly, recorded comments from the users including the hospital management, the catering staff, the nurses and the patients. This video program was then used successfully to sell many more installations. The video was particularly useful not only in convincing the managements of the benefits of the product, but also in defusing any criticism from the staff. Because they saw their counterparts enthusing about the system on the video (and the important thing here was that the people shown were real nurses and staff of named hospitals, *not* actors) the decision to change was not seen by the staff as a decision handed down from above.

It is not surprising that drug companies have been one of the biggest users of AV in selling. They have used, and continue to use, all the different media, ranging from filmstrip to computer-controlled videodisk. The main use has been for the presentation of case histories and clinical data, the kind of information that it would be difficult for the salesperson to remember precisely when there is a continually changing and wide range of products. The same thinking can be applied to any technical product that cannot be demonstrated on the spot.

Thus, at the other extreme, vendors of large machinery can use film or video to show their product in action. Even if the customer is able to see the product itself, for example a large piece of earth-moving machinery, it may not be practical to show it in operation in the way that the customer might use it. AV can help complete the picture for the customer.

The selling of services, such as consulting engineering, is most efficiently and convincingly done with AV help. This is not only because the use of slides or video allows the consultant to show the tangible results of his work, which may be several thousand miles away from that required by his new potential customer, but also because it permits the use of a well-written and well-presented summary that consistently gives the same quality message.

Selling benefits is what all sales people are taught to do. Encyclopaedia Britannica frequently use AV support at the individual home sale level to put over the benefits case for their product. This is an example of a case where the features of the product could be demonstrated directly, but the argument is better made with AV support. Sometimes AV helps the other way around; it can underpin both types of presentation.

Choice of equipment

The need to ensure minimum overall sales time also dictates the choice of medium for the person-to-person presentation. Usually the salesperson must visit the customer, so the medium chosen must use robust lightweight equipment, and must be able to give a show in an office or desktop environment.

Figure 8.1 Portable filmstrip units give a bright image and are lightweight. (Photo Edric Audio Visual Ltd)

Fortunately there are a number of self-contained portable devices available which use filmstrip, Super 8 movie, videotape, slides and computers.

The traditional media of filmstrip and Super 8 may still be of interest to managers of large sales forces looking for economy and impact. Filmstrip units can give large bright single images, and are significantly cheaper and lighter than their slide counterparts. However, they are not as flexible, and filmstrip as a medium is most economical where a large number of copies are needed of a fixed program.

Super 8 movie-based devices are the lightest weight and give the biggest picture for portable shows which need movement. The film is normally held in an endless loop cartridge, so once it has been started it must be run through to the end of the show.

However, movie is giving way to video: despite the smaller picture and, in general, higher price, users seem to be happier with the now familiar videocassette. The most portable units are based on the VHS cassette. An interesting variation is a new unit introduced by Sony (at the time of writing only available on the American National Television Standards Committee (NTSC) market) which combines a Betamax videocassette player with a small video projector, thus making the whole portable package suitable for small groups. In fact, users of the device have to be careful to ensure both the right lighting conditions and the use of a high-gain screen; in some ways this disqualifies it from this application, in which all equipment should be able to operate without the need for a special environment.

Slide-based units are by far the most flexible, and allow the smallest of businesses to use AV as an economical selling tool. Most such units are available in record/replay versions, so it is quite practical for users to make up their own simple programs. While such an approach is wholly justified if the business concerned is a small one which wants a lot of short programs for use on only one or two machines, any fleet operator should get the sound commentary, at least, professionally prepared.

A final possibility is to use a computer unit to present a sequence of computer graphics. This approach is only useful to a user who does not require real photographs of products, but simply wants a sequence of text graphics. This greatly limits its use so it must really be considered a gimmick, possibly suitable in individual sales campaigns, but not for continuous use.

Managers of large sales forces often use AV to support an individual sales campaign. For example, they may find it most effective if they use it two or three times a year, and their sales force ensures that all regular customers get to see the show, which might promote the launch of a new product or introduction of a new service. Such occasional fleet users may decide to rent the equipment for the campaign period only.

The direct mail video

So far this chapter has been mainly concerned with the use of AV as a support to an individual sales presentation. The huge population of videocassette recorders which now exists opens up another possibility.

Several companies have successfully used video as part of a direct mail campaign. This is now practical in many countries because nearly 80 per cent of the videos are VHS format and there is very wide home ownership of video machines within the target audience, as well as an increasing presence of video equipment in offices.

Two examples from the UK give an idea of the method in action. The Thames Water Authority, who are responsible for water and drainage in the south-east of England, offered all their customers a videocassette at well below retail price. By clipping a coupon they could receive a standard three-hour cassette; however, at the beginning there was a film describing the activities of the Authority and how they affected the consumer. Those who received the cassette could reasonably be expected to watch the film at least once, before they used the cassette for other purposes. A neat piece of public relations, where the actual cost of distribution was self-liquidating.

The magazine *Industrial Marketing Digest* in the UK reports on case histories of successful industrial marketing, and most issues include a story related to the use of AV. One 1985 issue reports the case of the Lovell construction group who successfully used direct-mail video to generate more business. They perceived that their company was not attracting as many enquiries as it should be, partly because the possible customers had an incorrect appreciation of Lovell's size and capabilities. They commissioned a video film, featuring a well known presenter/interviewer, that allowed customers and architects to explain, via interview answers, exactly how Lovell had helped them. The whole fifteen-minute production built up a picture of a company able to tackle a range of work with special skills and great economy.

Four hundred of the cassettes were sent out by direct mail to prospects known to be worthwhile

by Lovell, but the main distribution (another 1200 or so copies) required potential customers to phone Lovell in response to trade advertising. The results, measured in actual new business, were so good that Lovell went on to make more of these types of program aimed at promoting particular parts of their business.

Some interesting points about this exercise are:

☐ The actual cost of the cassette is only about double that of an expensively printed brochure (when applied to the relatively short print runs normal to industrial sales).

☐ This approach can *only* work if the video program itself is made to the highest professional standards. However, with careful planning this need not be excessive. The first Lovell program cost $23 000 including the fairly substantial fee paid to the presenter.

☐ While those who receive the cassette at their own request can be counted on to see it, there may be some doubt as to whether those sent it unsolicited would view it. In practice it seems they do. Expensive literature often goes straight into the waste basket but items of a perceived intrinsic value, like a videocassette, do not. They are often viewed out of curiosity (even if there is a later intention of using them for something else!).

Point-of-sale audio-visual

There are two kinds of point-of-sale AV. Either it is intended to augment the efforts of the sales assistant, or it is intended to be seen by passers-by, who, it is hoped, will be moved to buy. In the former case the rules are the same as for a person-to-person presentation, but because there is no need to move the equipment, bigger pictures can be used, e.g. via 26-inch video monitors.

The second case consists of passive and active displays. Either passers-by see whatever is put in front of them, or they have a means of expressing a choice. In the case of the conventional linear program it is important:

☐ that shows are short. More reference to this in walk-past shows is made in Chapter 9;

☐ that straight TV commercial shows are avoided. Shows should inform and, if relevant, demonstrate;

☐ that any equipment used is designed for continuous operation.

The aim of point-of-sale AV is to increase sales by showing sales material in a way that cannot otherwise be achieved. This is best illustrated by the following examples:

☐ The demonstration of do-it-yourself tools and materials in action, probably by video.

Figure 8.2 A striking rear-projection video unit for display use. It has a 55-inch screen. (Photo Cameron Communications Ltd)

Good-looking display cabinets are available which give random access to 80 or 160 slides. They make a reasonable-cost, high-quality presentation catalogue with individual images available within two or three seconds of selection. Slides give excellent quality and allow inexpensive updating. Technically more than 160 slides could be used, but for this type of application 160 is a realistic practical limit because an additional projector head is needed for every 80 slides.

☐ The explanation of how microwave cooking differs from conventional cooking, and how this can affect both lifestyle and the variety of food available. This would be a promotion for the whole idea of microwave cooking, rather than a promotion for a specific product, and is best done on video.

☐ The illustration of furnishing and decorative fabrics, such as curtains and wall coverings, in use. Here the sample book is augmented by brightly coloured pictures of room settings using the fabrics. This is best achieved by slides on a screen, say, 90 cm × 60 cm (3 ft × 2 ft).

☐ The selling of houses. Visitors to a real-estate agent can see a selection of houses by random access to color slides.

☐ The selling of vacations. A video presentation system shows a wide choice of programs on different vacation venues. Price and availability information held on a computer can be superimposed.

Random access is particularly suitable for point-of-sale work, either to allow customers to make their own choice, or to allow the sales assistant to show customers several alternatives on a quick selection basis.

Figure 8.3 This compact display uses slide projection to achieve a bright 180 cm × 60 cm (6 ft × 2 ft) image, while occupying only 0.6 sq m (6 sq ft) of department store floor space. (Photo Associated Images Inc.)

Figure 8.4 A random-access slide projection system suitable for the point-of-sale display of services, property, furnishing fabrics and other items requiring a high-quality still image. (Photo Diekhoff GmbH)

Random access on video tape is possible, but it is very slow and takes many seconds or even minutes. If it has to be done it is best to have a caption generator that can keep an image on the screen while the videotape recorder is searching. In general, tape is best reserved for conventional linear programs.

The best system for the random access of both moving-image sequences and for large numbers of still images is the use of videodisks. The videodisk medium is particularly economical for multi-outlet displays, but the initial production cost of the disk may deter some smaller users. Random access on the videodisk is very fast; in a well designed presentation it is often instantaneous, and never more than two or three seconds.

Systems which are both videodisk and computer controlled can use 'touch screen' technology to simplify both the customer's choice and the technical installation.

Figure 8.5 Out of hours window shoppers use an interactive videodisk display at the British Telecom Shop in Ealing, England. (Photo Convergent Communications Ltd)

Figure 8.6 The Indian Tourist Board using a computer-controlled random-access projector in an in-store tourism exhibition.

Displays using these technologies can be more than simple random access. If there is a situation where one choice leads to another – or, better, where the program segment shown depends on information provided by the customer – the display can properly be called an interactive display. The subject of interactive audio-visual is discussed further in Chapter 12.

The shop window

Retailers often ask for AV shows suitable for putting in shop windows, or similar public displays in shopping malls. There are two problems here.

The first is the nature of the program. Usually it has to be one that survives purely on its visual merits, because often the site does not permit the use of sound. The program should be short, or at least be in short sections, because it must quickly make sense to anyone passing it.

The second problem is light. Daylight, and particularly direct sunlight, is so much brighter than artificial light that it washes out most AV displays. Therefore displays using large projection screens can only be operated at night, or in an artificially illuminated shopping mall. Provided they are set back in a window, video monitors are usable, but a single monitor does not make much of a display. Thus, multi-screen video or video wall technique is emerging as a possible medium for this application. These systems are described in Chapter 19.

Figure 8.7 Olympus Sports make use of an eyecatching video wall in the window of their Oxford Street store in London, England.

Sales meetings

Here a distinction is made between sales conferences and sales meetings. Events designed to motivate a sales force were discussed in Chapter 6. A different type of event is the meeting where the intended outcome is a direct sale to individual members of the audience.

The first thing to remember is that it is very difficult to deal with a large number of people. Once the audience has seen the presentation, how can the presenters get all of them to buy?

There have been successful sales meetings to large audiences, especially in the travel business, but only when there has been adequate staff present to process the audience at the end of the meeting.

Figure 8.8 The showroom of Haagtechno (the Dutch importer of Panasonic Technics products) uses multi-image to support sales presentations.

The sales meetings using AV that are usually unqualified successes are those where the audience consists of a group of people from a single customer. This method of selling is becoming of increasing importance to all companies with a sophisticated product or service to offer. This type of meeting takes quite a lot of setting up. Precisely because of this, it is important that they get results. AV will help ensure this by:

☐ giving a proper structure to the meeting;

☐ giving a sense of occasion;

☐ minimizing the overall executive time spent;

☐ improving the communication of the basic message.

The slide/sound or multi-image medium is usually best for these types of meeting or presentation. It allows maximum impact for the recorded part, and the highest quality for the

Figure 8.9 A.E. Le Page, Canada's largest real-estate company, uses AV to sell office accommodation. This is the information center at College Park, Toronto.

visual-aid sections. It only requires one set of equipment, and presentations can easily include customized visuals. Examples of successful users of this technique are to be found in all industries, but especially those selling capital projects and equipment (e.g. telephone systems builders and building contractors), those selling services (e.g. engineering and management consultants) and those selling property (e.g. the renting out of a prestige building, or the selling of a new development).

New methods of slide production allow high-quality images to be prepared at minimum cost and, probably more important, in minimum time. It is also possible to give presentations based on video, but in quality terms this is only suitable for very small audiences. The recorded section of the presentation can be on videotape, and the customized material can be generated and stored in a small microcomputer.

The correct way to conduct this kind of sales meeting is to use a presentation room (see Chapter 11).

Public relations

Audio-visual is a powerful public relations tool. Some examples are described below.

A company already has a presentation room. It makes a point of having a good 'house show', not only for employees and customers but also for the general public, e.g. school parties, local residents and special-interest groups. Some users develop special programs for such audiences. An exceptional example of this is to be seen at ASEA's headquarters in Vaesteras, Sweden. This giant electrical company has a company visitors' center

Figure 8.10 Siemens use multivision in an elegant display for waiting customers at their offices in The Hague, Holland. (Photo International Film Services BV)

called ASEA Forum where considerable use is made of all kinds of AV techniques. The center is used both to support the selling effort and, particularly, for PR.

A company is invited to sponsor an exhibit in a museum or other public visitors' center. An AV presentation may not only meet the requirement of the museum, but allow the sponsor a method of putting its name in front of the audience in a way that is most acceptable. For example the Boeing Aircraft Company sponsored the exhibit, and a specially produced multi-image show, on the subject of flight, at the Singapore Science Center. At Britain's National Motor Museum several well-known companies such as Lucas, Alcan, Kennings and Ford have sponsored exhibits that incorporated AV. An AV show can also help to give structure and a sense of occasion to press conferences and similar events.

The examples of AV use described above are open to all sizes of organization. Larger companies with household names can also consider sponsoring films and video programs, with a view to either having them shown on the public broadcasting network or, more realistically, having them distributed to specialist audiences. In this case it is important to plan how the film is to be distributed at the time it is commissioned, because this may well affect how a particular subject is

Figure 8.11 A compact video presentation unit in use in an automobile showroom. (Photo Bell & Howell Visual Communications Division)

tackled. Distribution of this type of film is best handled, at least in part, by specialist library firms who have both access to the audience, and a streamlined system for the issue, recovery and maintenance of the film copies and videocassettes.

Chapter 9

Audio-visual in exhibitions

Trade exhibitions are a natural use of audio-visual. In fact, trade exhibitions in Europe were, for a number of years, the main application of some audio-visual media. There are two opposing attitudes to the use of AV on exhibition stands; one of delight at how effective the medium can be, the other of disillusionment because an expensively produced show has had little effect.

There are so many positive things that AV can contribute to a trade exhibition stand. Here are just a few examples.

Making more effective use of the visitor's time
Most exhibition visitors have limited time to absorb the content of a particular stand. An AV show may be used to summarize what is on offer. A different approach is needed depending on whether the show is aimed at existing customers paying their routine visit to a supplier's stand, or at potential new customers seeing the company's product range for the first time.

Attracting more visitors on to the stand
Although AV can be used for several different purposes in exhibitions, a particular show can seldom fulfil more than one objective. It is possible to use AV as an attraction to entice people on to the stand. However, it is unlikely that this kind of show would also be able to meet the effective use of time objective.

Solving a language problem
AV shows can have multilingual commentaries. This is only worth doing if the environment allows the visitor to see the show in comfort. It is more important that the visual sequence includes interpretation clues to aid foreign visitors.

Contributing to more effective selling on the stand
Most trade exhibitions are short. Anything that helps make more effective use of staff on the stand makes a better return on the exhibition investment. For example, at busy times visitors waiting to see a particular salesperson can watch an AV show. The right show ensures that when the salesperson is free, the customer is well disposed to the exhibitor, and has already received some of the sales information.

Providing better public relations
Often an exhibition stand is one of the few places where an industrial company is on public display. It can be just as important for the company to look good in the eyes of third parties, as to make sales from the stand. An AV show can help the PR image of a company in the eyes of suppliers, downstream customers, the public, and direct customers. In some cases the AV can help entertain the public, who may have no need to take up the valuable time of sales staff, while serious visitors can be personally attended to.

There are many positive reasons for using AV techniques at trade exhibitions, but there are also some negative aspects. Negative results reported by some users are summarized in the following comments:

> 'The picture looks all washed out, far from being an AV spectacular, the exhibit looks weak.'

> 'We had lots of people looking at the show, and our stand seemed to be the hit of the exhibition, but we made no extra sales and got no more leads.'

> 'We had to turn the AV off because it was preventing our sales staff closing sales with customers.'

> 'We have this great AV that lasts 15 minutes and cost thousands of dollars. But on the stand nobody watched it for more than a minute or two and they usually missed the bit about the new product.'

> 'The AV was a pain. The sales manager is the only person who knows how to work it properly, and

he was too busy. The fuse kept blowing, and the projector was in the way of the coffee machine.'

These complaints betray either a lack of commitment to making the AV element do its job, or more excusably, an ignorance of the factors that must be taken into account when using AV in the special environment of the trade exhibition.

Because any company or organization participating in a trade exhibition is hoping to make a reasonable return for the investment in stand space (whether the return is measured in money terms, or in indirect benefit terms), it is important that those using AV do not fall into unnecessary and wasteful traps. How, then, is disappointment to be avoided? The key to this is understanding the nature of the problems:

☐ Any stand is competing with many others. Will the potential audience have time to see a complete AV show?

☐ There is no way to control *when* a visitor will arrive on the stand. An AV show half-way through is more likely to make visitors move on than attract them.

☐ Can the exhibition stand be designed so that the AV show can be seen and heard properly, and yet not interfere with other activities and displays on the stand?

If the objectives of the exhibition stand are fully understood, it is easier to see how the problems can be solved. There are specific points that the exhibition manager must consider.

If portable audio-visual equipment is to be used, then it must be used in the same way as it is on the road. This means that sales staff must be in attendance to introduce the show and to talk to the visitors. Therefore the AV units must be sited to allow individual presentation. The exhibition stand must become an extension of the salesman's territory. If this condition cannot be met, it is much better not to use the AV program at exhibitions.

If a conventional AV show, designed for group communication, is to be shown, there must be a semi-enclosed viewing area. The exhibitor must make sure, in this case, that he can hold people until the show starts and that there are enough staff on hand to talk to each group after the show. A theater-type show is really only suitable if the exhibitor has a highly targeted audience, or the show is fulfilling a PR role. An example of the first case is a manufacturer of telephone exchange equipment selling in a country where the telephone system is a state monopoly. In that case there is only one customer, so the show can be targeted at the people who matter in the buying organization. An example of AV fulfilling a PR role is a public-image show put on by a consumer

Figure 9.1 Exhibition shows with a story must have a proper viewing environment. (Photo Electrosonic GmbH)

goods manufacturer. Here the show can be considered as part of the overall PR effort, and its success measured by size of audience and audience appreciation.

If the controlled environment of a theater is not possible, and often it is not, an effort must be made to integrate the AV into the exhibit fabric, and to ensure that viewing conditions are correct. This means paying attention to details of staging, such as avoiding direct light on the screen, using back projection, providing a leaning wall or barrier for viewers, and ensuring the correct screen height.

A show on an open stand must be completely different from a 'house' or sales meeting show. It must be *short*, ideally two minutes long, but no longer than five minutes. It must be written in a circular fashion, so that it makes sense to visitors within a few seconds of joining in and ensures that they will see the whole sequence through, regardless of where they started.

Two important ideas emerge from these points. Firstly, an existing program investment can be re-used. If a company has already commissioned a show from a producer, e.g. a house show, a sales meeting show, or any other show intended for a controlled audience, using the same producer and same basic material it should be possible to create a special exhibition version which is shorter and sharper. This kind of show can be very successful.

The second idea is to use AV as a 'come on', an invitation on to the exhibition stand. In this case the AV show must be integrated into the decor. In fact, it becomes much more a part of the stand design than a detailed communication method.

This approach requires short, simple and bold program material and is one of the best applications of multi-screen technique. It is particularly suitable for general public exhibitions.

Choice of medium

In principle all the different AV media can be used on exhibition stands, but there are some guidelines which should be followed.

Figure 9.2 An unusual screen format adds interest, especially when, as here, the images are bright. (Photo courtesy of Vickers PLC)

A TV set is just a TV set and its use as a display may, by its very familiarity, devalue a show on an exhibition stand. The exhibition designer must take care to 'design in' video. Some of the recently available rear-projection TV sets may be more suitable than a conventional set. Better still, if space permits it, is the use of video projection with high-gain rear-projection screens. These screens are available in sizes equivalent to a 45-inch or 67-inch TV set.

Individual communication units, such as the portable single slide/sound units, or compact video units, should not be used as display devices. Instead, an environment must be created in which they can be properly used by sales staff.

The multi-image technique using slides, whether on one or 100 screens, is one of the most spectacular media for exhibitions, and is easy to adapt to special needs. Back projection is the only satisfactory method of presenting optical projection on an open stand, and it is best to keep individual image areas small and bright. This simply emphasizes the point that the multi-image show made for a sales conference is unlikely to be of any use on an exhibition stand. On the other hand production costs of exhibition multi-image *should* be low in relation to effective use of the stand. Shows should be short and based heavily on existing high-quality visual material. If a show producer offers a 15-minute show using only specially created material for an exhibition stand, either he knows little about trade shows or imagines he has a gullible customer. Equally, the exhibitor should specify very simple objectives for these shows and not complicate the message by asking them to do too much.

Figure 9.3 A single video monitor on an exhibition stand can look insignificant. The UK Department of Energy uses a clever five-screen video production on its stand at offshore technology exhibitions.

Figure 9.4 The adverse effect of ambient light can be minimized by setting the screen back. Here the designer has used the simple trick of side mirrors to make the compact AV display look bigger.

Multi-screen video is an exciting and fast-moving medium; however it is a great deal more expensive than multi-image. An exhibitor would usually need to amortize the cost of a multi-screen video show over several exhibitions to justify it. Multi-screen video is described in detail in Chapter 19.

The prestige exhibition stand, especially one at a show open to the general public, can benefit from a multi-media approach, where a synchronized sound and light technique is combined with projection. There is no limit to the possible size or complexity of such shows, so it is entirely up to the sponsor to decide what is justified based on his own criterion.

An attractive idea is to have displays that offer the visitor random access to different images or different programs. The problem here is that only a limited number of people can use such displays in a crowded exhibition. They therefore tend to be of greater value in permanent or long-running exhibitions, or in specialist trade shows where the exhibitor can justify a high cost per serious visitor. The possibilities are described in detail in Chapter 12.

Multi-language displays are of special benefit at international exhibitions. These can use either multilingual recordings, multilingual graphics on the screen, or both. They are, however, only a support for the selling effort in the customer's language, not a substitute for it.

Staging

The main practical problem in using AV on an exhibition stand is the less than ideal circumstances of installation. A wise exhibition manager takes steps to insulate himself from the problem by appointing a competent staging house or rental company to both install the equipment, and to ensure its proper running throughout the course of the exhibition. Even if the exhibitor owns the equipment, it may be best to contract out its installation and maintenance unless there are staff whose sole job is AV support.

It is important to allow enough time for installation. This means that other trades, especially the electricians providing power, must finish early enough to ensure that the AV contractor has time to complete the installation, which can only be started after everyone else has finished.

One common problem is adequate power. Simply having enough power sockets is not always enough. If only one or two video units are involved, there is unlikely to be a problem, but if twenty slide projectors are to be installed the story is different. In theory the power may be sufficient to run all the projectors, but in practice the switch-on surge may trip circuit breakers, or a high-impedance supply may result in mutual interference between projectors. These problems are entirely avoidable if the correct power supplies are specified and installed.

It is a false economy to skimp on the space provided for the AV. Either it forms an important part of the exhibition stand, in which case it should be properly staged, or it does not, in which case it is best omitted. Therefore, the rear projection area should *not* also serve as the coffee room or catalogue store. Equipment *must* be accessible for service without inconveniencing visitors.

It is quite possible to arrange for AV shows to be staged in overseas countries. The question as to whether it is best to take the equipment from the exhibitor's country or to rent it locally will depend entirely on circumstances, and should be reviewed with the staging company. In general, simple shows using standard equipment configurations are best done on a local rental basis, but only if the program material is sent to the rental company well ahead of the exhibition date. This allows the show to be tested on the actual equipment it will run on. Complex shows often need equipment and staging from the exhibit's country of origin.

Chapter 10

Audio-visual in museums and visitors' centers

The old idea of a museum being a place where dusty archaeological remains are viewed in an equally dusty cabinet is, fortunately, now out of date. Today there are museums on subjects as diverse as geology, tennis, science, horse racing, natural history and beer. There are also an increasing number of sites which cannot be called museums, but which have many of the same requirements, both in terms of what the visitor expects, and in their day-to-day management. It is easiest to refer to these as visitors' centers. They may vary from an exhibition gallery attached to a tourist site such as a National Park, to a permanent exhibition at a company headquarters. A traditional museum can serve as:

□ a place of study for scholars of the museum's subject;

□ an archive, where the storage of antiquities or artefacts is more important than their display;

□ an educational institution;

□ a place where items are on display and an effort is made to interpret their importance and relevance to visitors.

Major institutional museums will serve all four objectives, but the smaller museum and all visitors' centers are only attempting to display or interpret exhibits. It is in this area that audio-visual can make a significant contribution, both to the visitor's enjoyment and appreciation of the subject, and to the way in which visitor sites are managed.

Audio-visual techniques can assist the interpretive function in many ways, for example:

□ solving language difficulties;

□ creating an atmosphere;

□ changing the environment;

□ allowing comparison with outside material;

□ setting a context;

□ providing historical background;

□ developing an idea;

□ introducing a subject;

□ demonstrating a process.

They can also help in the management of visitors. In some cases AV can make the difference between a particular site being a worthwhile place to visit, and an unsatisfactory visitor experience.

Museums and visitors' centers represent one of the most interesting and potentially successful applications of AV techniques. The fact that they are permanent installations enables the best value to be extracted from a program investment, but this permanency also introduces some special problems and disciplines. In principle all the AV media have a place. This chapter reviews some of the points unique to the permanent installation.

Show environments

AV can either be shown in special auditoria or can be integrated into an exhibit. In the first case special attention must be paid to the ease of exit and entry to the auditorium. The question of whether the audience will stand, sit or have leaning rails depends on the length of the show. Shows longer than twelve minutes should have seating.

An auditorium is only used where the show stands on its own. Many applications require the AV to be built in to the exhibit fabric. If this is done, shows must be *short*. (See also Chapter 9.) It is also important for the show to be properly 'designed in'. It is not good enough to install a piece of equipment like a TV set in the middle of a gallery, it must be designed in so that the screen forms a logical part of the display.

Figure 10.1 The Theatre of the Eagles at Caernarvon Castle in Wales. A mixed multivision and movie show with additional lighting effects. (Photo Abram PR Ltd)

Separate auditoria have the advantage that lighting conditions can be controlled to ensure optimum viewing of the show, but the separate show is only one possibility; often an integrated show is more appropriate. In this case special care must be taken to ensure that viewing conditions are correct. Particular attention must be paid to prevent front lighting from falling directly on the screens, and to make sure that the screen size is suitable for the prevailing ambient light conditions.

Figure 10.2 Video screens built in to an exhibit set. Here two rear-projection screens (one simulating a pond) and one monitor are built in to an exhibit on natural history subjects. (British Pavilion EXPO 85)

Choice of media

The choice of audio-visual medium will be made on the usual bases. However, in permanent exhibition work:

☐ 16 mm movie can be used, either as endless loops or with automatic rewind equipment. However, its running cost is high, and it requires regular maintenance. It is best reserved for use in small auditoria, rather than within exhibits.

☐ Video is reliable, but the familiarity of a TV set can devalue the show. Clever designing in can help overcome the problem. Front-projection video can only be used in small auditoria, but rear-projection video using special high-gain screens up to about 1.4 m (4.5 ft) wide is entirely satisfactory as a built-in element.

☐ Slides are the ideal visitor or single-site medium. They allow shows, both single-screen and multi-image, to be made that are unique to the venue. These shows can have a great impact on the audience, and are easily kept up to date.

☐ The permanent exhibition allows the use of the synchronized sound and light or 'Son et Lumière' technique. This can bring a display to life. The system may be as simple as lighting up a few transparency boxes in sequence, or as complex as having fully animated figures with programmed theatrical lighting.

☐ The media can be mixed. A frequently successful example is where a slide program accompanies the description of an animated map or diagram. In principle, it is possible to automatically program the activities of any device that can be directly, or indirectly, controlled electrically.

Figure 10.3 One of four audio-visual theaters at the Citadel Historic Visitors' Centre, Halifax, Nova Scotia. Each show describes one century in the Citadel's history and uses a mixture of slides and son et lumière technique. Shows are in English and French. (Photo Multivision Electrosonic Ltd)

The auto-present concept

Some first-time users of AV in permanent displays do not appreciate the difference between running an AV program as part of a person-to-person presentation, and running one in a fixed installation on a day-in, day-out basis. Equipment that is suitable for the one may be quite inadequate for the other.

Ideally fixed shows should run fully automatically at the touch of a button. There can be no operator. This is where the idea of the auto-present show is introduced. Its attributes are best illustrated by the example of a permanent exhibition with a small auditorium show based on the multi-image technique. It has houselights in the auditorium and a set of motorized drapes across the screen. The auto-present method of operation works as follows:

☐ A *single button* is pressed. This causes the following:
○ the projector power switches on;
○ the tape starts.

☐ The show automatically runs from the tape, which can also program auxiliary devices. Therefore it is used to:
○ fade down the houselights;
○ open the screen drapes;
○ run the main multi-image show.

☐ At the end of the show the sequence is also fully automatic. The tape is programmed to:
○ close the screen drapes;
○ fade up the houselights;
○ instruct all projectors to return to zero or 'home';
○ rewind the tape to the beginning;
○ when all projectors are 'home', switch off the projector power.

The same principles can be applied to any automatic AV show, whatever the medium being used. The video media are easy to auto-present. Slide/sound and multi-image are easy only if the right equipment is installed. Movie film is also auto-presentable, but does require special projectors or projector attachments.

Where a mechanical device such as a slide projector is used in an auto-present system, it is essential that the homing procedure *independently* checks that the device is in the zero position. For this reason the projectors used must have zero-position sensing microswitches, and the control equipment must be of a kind that uses them. The same principles apply to other devices, such as turntables, that can take up several positions. It is easy to see that continuous shows can be given by allowing the system to restart itself, but only when it has checked that all equipment is properly reset.

A simple addition is to add a clock system that runs the show automatically at preset intervals or at preset times of day. An additional feature is a count-down clock which shows the audience how long they must wait before the next show.

It is vital that the proper running of these shows be monitored, but one of their benefits is supposed to be unattended operation. The best thing is to have a routine inspection. One helpful feature for shows based on projection is an automatic lamp changer. This device is available for current models of automatic slide projectors.

The enemies of reliable operation of AV equipment are *dirt, dust* and *excessive heat*. It can be a worthwhile investment to provide a special environment for the equipment. For example, projection rooms should have their own filtered air supply. Video recorders can be centrally sited in a clean, air-conditioned, or at least cool, room, and slide projectors in individual exhibits can have their own protective box with positive pressure filtered air supply. The correct installation of AV equipment can make the difference between negligible or appreciable maintenance costs, and can increase equipment life by years.

It is very important that equipment be *accessible* for maintenance. Designing an AV show without providing proper space and access for equipment is like designing a restaurant without a proper kitchen.

The role of the videodisk

Until recently it could be taken for granted that any simple video display in an exhibition or museum would be run from videotape, usually U-Format, but more recently often VHS or Betamax. However, videotape does have some

Figure 10.4 The video replay system at the Guinness World of Records Exhibition is based on the use of U-matic videotape players. Although reliable, they do require regular maintenance. Provided programs do not have to be changed too often this type of installation would now be better done with videodisk.

disadvantages as a medium when used on a continuous day-in day-out basis. The following points should be considered:

☐ Continuously running shows can only be achieved using two machines in tandem. Otherwise there is a significant rewind interval, which means a blank screen during rewind if an interval caption or program to cover the interval is not provided.

☐ To minimize wear on the cassette and player of a continuously running show, the program must be repeated as many times as possible down the cassette. For example, a five-minute video sequence should be recorded at least six times onto a thirty-minute cassette. This puts up the cost of dubbing the cassette.

☐ Tapes wear out after a few months and must be replaced.

☐ Picture quality is seriously degraded by dirty tape heads. Regular tape head cleaning maintenance is therefore required.

☐ Worst of all, the tape heads actually wear out. They may need replacing every one or two years, which is expensive.

☐ Museum and exhibition designers like systems that allow visitor choice, and can give good still-frame performance. Videotape is a poor performer for still-frame use, and is very slow in any random access application.

The advent of the industrial videodisk player (described in more detail in Chapter 19) is now transforming the way in which exhibition video is done, because it overcomes many of the disadvantages of tape.

The first reaction of a museum curator to the use of videodisk might well be one of horror at the cost of creating a videodisk. However, if overall costs are examined, it can often be shown that videodisk is actually less expensive than videotape, as well as introducing operational advantages.

The simple industrial videodisk players are less costly than industrial-standard videotape players. Even if only one year's operation is taken into account, the elimination of both routine maintenance and replacement cassette dubbing yields a worthwhile saving.

However, the approach really only makes sense when considered on a system basis. An example best demonstrates the principle. In the British Pavilion at EXPO 85 in Tsukuba, Japan, it was necessary to run fifteen separate video sequences. Each was to run continuously, fourteen hours a day for six months and each sequence was about three minutes long. The conventional solution would have been to use fifteen U-Format machines with, for example, thirty-minute cassettes with multiple program repeats. The rewind interval would have had to be tolerated, or dealt with by an interval logo requiring a lot of extra equipment.

It was actually less expensive to use fifteen videodisk players. These were all sited centrally for ease of maintenance (although over the six-month period none was actually required) and controlled by a microprocessor-based multi-show controller. This device simply ensures that each

player only played the sector of the disk appropriate to its exhibit. The total playing time of all fifteen video segments was about 58 minutes, so it was necessary to originate a double-sided disk, because one disk side holds a maximum of 36 minutes of program material in the Laser disk CAV format.

A videodisk allows two sound tracks with the video program. In the example most of the video sequences only required a single sound track, or were mute. However, several 'wild' sound tracks were needed to accompany other, unrelated, exhibits – for example the sound of children at play to accompany a schools exhibit, or the sounds of wind and waves to accompany a North Sea oil exhibit. It made economic and practical sense to put these sounds on the unused audio sections of the same disk. In this way a high-reliability, high-quality sound replay system was provided at a very small extra cost.

At Tsukuba the video system was required to run continuously, but this is not a necessary condition of this type of system. Each of the individual video programs could have been either continous-running or push-button-start. Videodisk has the added advantage that the screen need never be blank. It can maintain a still frame for months without damage to the disk, so while waiting for a visitor the disk can display a title or instruction image.

The same thinking can be applied to a museum or visitors' center. A craft museum might wish to use audio and video technique to:

☐ Run a continous three-minute video program to describe the museum as a whole.

☐ Run three other continuous video programs, each of two minutes, and each accompanying a particular exhibit. They might show the machine/artefact/tool being used.

☐ Have an exhibit giving a push-button selection of ten one-minute video sequences, each demonstrating a different facet of the use or making of objects related to the museum subject, but which for some reason cannot be exhibited themselves.

☐ Have an exhibit giving a push-button selection of 100 (or 1000!) still images.

☐ Have one or more simple interactive displays, testing the knowledge of visitors about particular subjects or, at the end of the museum tour, testing what they have learnt in the museum.

☐ Run several continuous sound tracks to accompany other exhibits, for example the sound of birdsong, of people in a reconstructed dwelling or shop, or of machines.

Figure 10.5 The Philips VP410 videodisk player is suitable for many standard video replay applications in museums and visitors' centers.

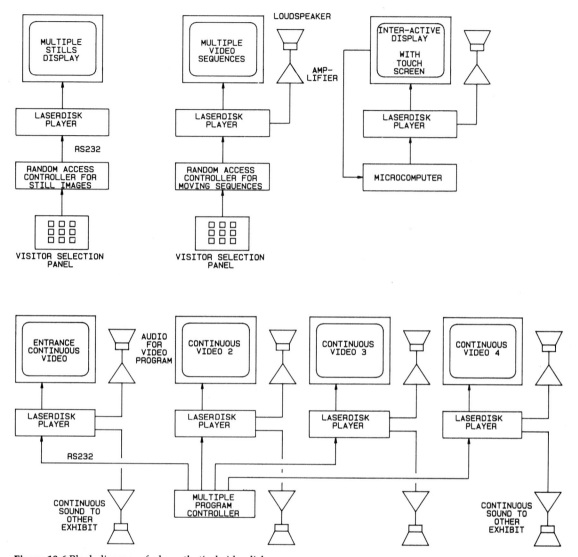

Figure 10.6 Block diagram of a hypothetical videodisk replay system for eight different exhibits in a small museum.

With careful planning all these items could easily be fitted on a single disk side. If the project were handled conventionally (by use of videotape, sound cartridge machines, etc.) the preparation work would still have to be done, so the only extra work is originating the disk. The cost of this is quickly saved in lower system capital cost and lower running costs. The benefits also include better picture quality and the ability to achieve fast-responding interactive displays that cannot be achieved by any other method.

Figure 10.6 shows the equipment needed to achieve the hypothetical example. Clearly it would be up to the exhibit designer to integrate the video screens into the exhibit fabric, and although video monitors are shown in the diagram there is no reason why some of the sequences could not be shown by projection.

Interactive displays

The word interactive is somewhat overused, and for museums it is really better to talk in terms of visitor-operated displays. Sometimes all the visitor is being asked to do is to express a choice, for example a choice between a number of video programs or a number of still images. These displays are very useful where a lot of facts must be available, almost on a catalogue basis. The possibilities are:

☐ The push-button random access of slides. Random-access slide projectors should be used where big bright images are needed, and where changes might be needed. One random-access projector can hold 80 slides, although it is possible to have several projectors working on the same screen to give a larger capacity. Four projectors, 320 slides is a practical maximum.

☐ The random access of still images on videodisk. The possibilities here are almost limitless because one disk alone can accommodate 54,000 still images. In both this case and that of slides, the simplest method of selection is to offer the visitor a decimal keypad, usually with a simple two or three digit display confirming the slide selected. It is also possible to have one button per selection, which can make for a better display.

Figure 10.8 The display of current sporting records, which can be updated daily, is done by a special computer display at the Guinness World of Records Exhibition in London, England.

☐ The random access of video sequences. Although this can be done from video tape, the slow response time and higher maintenance costs make it unsatisfactory compared with disk. A Laser disk can hold 36 minutes worth of material (PAL format) or 30 minutes (NTSC format), so provided the total run time of all segments does not exceed these maxima, all that is needed is a single disk player with a suitable selection panel. This would normally be a special control which permanently stored the sequence information. It could, for example, give the visitor the choice of ten two-minute sequences and five three-minute sequences. The actual durations are not important, but the total must not exceed the disk capacity. If it does, more players must be used.

☐ Another interesting possibility is to use a videodisk to give a much larger capacity of choice of programs using still pictures with audio. Special disk production techniques allow many hours of audio to accompany still pictures; the actual capacity depends on the mix of audio and picture. A choice of 100 two-minute programs, each containing a sequence of twenty slides, would be easily accommodated.

Figure 10.7 At the Guinness World of Records Exhibition in London, England, random-access slide projection is used for showing record-breaking photographs.

☐ Random access is not confined to slide or video images. Random-accessed computer data can be in

Figure 10.9 The use of thump-proof push-buttons is recommended even when the visitor is the Lord Mayor of Westminster. (Photo The London Experience)

graphic or, more usually (because, in general, real images are best presented on slide or video), in text form. Such displays are ideal for facts and records (e.g. sporting records) because they can be kept up to date on a daily basis if necessary. It is important, however, that the displays format the data presented in an attractive form, and that the amount of information on any one page is limited.

A truly interactive display is something more than a choice of images, and much more difficult to do well. Thus, simple random access should be thoroughly explored first, because it meets many of the display requirements of exhibitions. Then, when a display is required that demands the active *participation of* and *input from* the visitor, real interactive displays should be introduced. This complex and new subject is covered in Chapter 12.

Before leaving the subject of visitor-operated displays, a word about visitors' grubby fingers and propensity to destroy push-buttons. Although thought of as a 'small boy' problem, experience has shown that boys and girls of ages 5 to 85 are equally likely to damage the input end of such displays. Therefore, the following points should be considered:

☐ Touch screens are popular with many designers, but need regular cleaning. They also make all interactive displays seem the same, so some designers prefer a custom push-button panel that can be better related to the exhibit design. Buttons are less expensive too.

☐ Push-buttons should be chosen for their toughness. Those used by gaming machine manufacturers are a good choice. For really difficult cases, buttons designed for use on heavy machinery can be considered.

☐ If a museum or exhibition has a number of displays with the same visitor panel; it can pay to have a custom membrane keyboard made. These can have artwork of

any complexity, as many or as few buttons, indicators and displays as required, and are waterproof. Once the tooling has been done they are inexpensive. They are easy to keep clean, tamper-proof and economic to replace.

Audio systems for permanent exhibitions

Audio systems for museums and permanent exhibitions is one area where the rules are changing fast. Before defining technical solutions it is best to define the requirements of these installations. They vary enormously in terms of the resources needed to give the required results, and include the need for:

☐ Continuous sound effects to enhance an exhibit, for example the sound of a locomotive in a transport museum; the sounds of birds and animals in natural history exhibits; the sound of people in environmental or room reconstructions.

☐ Sound accompanying small-scale audio-visual exhibits. These include video, single-screen and multi-screen slide shows, and small-scale sound and light shows.

☐ Sound accompanying large-scale and theater-based audio-visual shows, for example, the larger multi-image shows, sound and light sequences and animated figure shows.

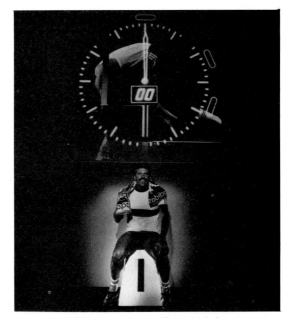

Figure 10.10 The 'superstars' gallery at Madame Tussauds in London, England, has high-quality continuous multi-track sound replayed from tandem multi-track tape decks.

☐ Exhibit commentaries delivered on a localized basis, sometimes in several languages.

☐ Acoustic guides, where visitors are given a device to carry round with them that allows them to have an individual commentary on each exhibit area, often with a choice of languages.

There is now a wide choice of storage medium for audio. The choice of which to use should be made on cost-effectiveness grounds, with particular emphasis on minimizing running costs and the need for maintenance. Ignoring, for the moment, the special requirements of the acoustic guide, the following choice of medium exists:

☐ Reel-to-reel tape, in quarter-inch, half-inch and one-inch widths, able to carry, typically, four, eight or sixteen parallel audio or control tracks.

☐ One-inch tape in an endless loop carried in a bin loop machine, giving typically eight, twelve, thirteen or sixteen parallel audio or control tracks.

☐ 16 mm movie film carrying either an optical or a magnetic sound track, or both.

☐ 35 mm movie film carrying either an optical sound track (mono or stereo) or up to four magnetic sound tracks; also 70 mm film able to carry up to six magnetic sound tracks.

☐ 35 mm magnetic film carried on a separate transport, able to carry up to eight sound tracks.

☐ Quarter-inch lubricated magnetic tape, carried in an endless loop cartridge. The best known is the NAB cartridge which is widely used in the broadcast industry, and is available from many vendors. An alternative is the Mackenzie cartridge which is less easy to use, but which is claimed to have a much longer service life. Other cartridges include the Mini 8 cartridge based on the old eight-track stereo automobile cartridge, now not recommended, and the Reditune™ cartridge developed for the background music industry.

☐ The compact cassette.

☐ The videocassette.

☐ The videodisk, both as a source for accompanying a video program, and just as a sound source.

☐ The compact audio disk.

☐ The hard magnetic disk (e.g. a Winchester disk store as used for computer memory).

☐ The recordable optical data disk.

☐ The solid-state audio store.

Some of these can be dismissed quickly. Sound carried on 16 mm or 35 mm film will only be used when there is a movie-based presentation, and even then there will be cases where the sound is carried on a separate tape.

The magnetic hard disk can store sound digitally, but has a relatively limited capacity. This particular application of the magnetic disk is not really the subject of any standards. Thus, it will only be used in special custom-built systems. One possible application is in the provision of a random access audio store. For example a 20-megabyte hard disk could store a choice of 170 fifteen-second speech messages, any one of which would be available within a fraction of a second. This method is valuable if messages are subject to frequent change. If they are not, a higher-capacity store with better sound quality could be achieved by using a videodisk, which is also easier to get made than a compact audio disk.

The compact audio disk can give superb stereo sound. However there has been a problem in getting these disks made in small quantities, so its use in individual exhibits is very limited. There is no problem in getting the disks to continually play only a section, so if part of an already available disk is required, it is easy to use. Alternatively, an exhibition could commission its own disk and use it in several different areas, along the lines previously suggested for the videodisk. If a compact audio disk must be synchronized to a display, it is technically possible, but does need special control equipment because a clock system which reads the disk position is the only practical way. In this respect, the videodisk is an easier proposition.

Figure 10.11 Many USA theme parks and museums use bin-loop machines for the reproduction of high-quality sound. (Photo California G & K Designs Inc.)

Most high-quality shows given in theaters or in large exhibition areas will continue to use reel-to-reel tape recorders for some time yet. This is because modern multi-track recorders are reliable and represent excellent value. Shows given on push-button or automatic-timed demand require automatic rewind units, that rewind the tape at the end of each show and precisely position it for the next start. They also form part of the auto-present system.

It might be thought that those exhibitions requiring high-quality audio on a continuous basis would have to use the bin loop system. Although bin loops are capable of high-quality sound, they are not made in large quantities. Therefore they tend to be very expensive. Servicing requires specialist knowledge and spare parts may not be easily available. For most users it can be better to use two reel-to-reel machines working in tandem, arranged so that the second machine starts up while the first rewinds. Provided the precise-start-positioning system is used, the change-over can be virtually imperceptible.

For small shows it is feasible to use the compact cassette as a sound source, but only if used in conjunction with a heavy-duty automatic cassette tape deck intended for exhibition use. It is critical that tape heads be regularly cleaned (as they should on reel-to-reel machines) and only cassettes of the highest mechanical and tape quality should be used. On the other hand, unsuitable tape formulations must be avoided; for example, few exhibition machines can benefit from the use of 'metal' cassette tape, and they will give worse results than standard formulation tapes unless the

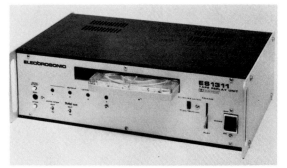

Figure 10.13 A cartridge tape deck designed to meet the needs of exhibitions. It plays four tracks from an NAB cartridge.

machine has been specially designed to take them. For show work the cassette deck needs an automatic rewind system working on a similar basis to that used on reel-to-reel systems. The endless version of the compact cassette should *not* be used for any serious museum or exhibition work; it is really only suitable for short-running applications, such as a two-day show.

The endless tape cartridge using quarter-inch lubricated tape has been the mainstay of continuous sound in museums and exhibitions. In principle, cartridge tape decks are mechanically very simple, so need little maintenance. In practice, the sound quality available from cartridges is limited, but acceptable for many purposes. Normally the tape speed (9.5 cm/s or 3¾ in/s) is used to obtain a reasonable compromise between sound quality and cartridge life. Tape heads tend to clog quickly, depending on the lubricant being used, so must be cleaned frequently. While there will continue to be some demand for cartridge systems, their use for museum and exhibition work will rapidly decline in favor of newer technology.

Videodisk can be used as a sound and picture source; so for some users, sounds which would have been carried on tape cartridge will now be carried on spare tracks of the videodisk. In the future the recordable optical disk may be used as a sound source, using digital techniques to store multi-channel sound. This may usurp the role of high-quality reel-to-reel machines in major shows. This type of system would have the advantage of mechanical simplicity and minimum maintenance. However, for the present, digital techniques have another manifestation which is of immediate importance and availability to museums.

This is the digital sound store. Most museum directors who use sound in their museums will

Figure 10.12 For the smaller exhibit the NAB tape cartridge has been the standard for continuous sound. In the future it may be supplanted by solid-state sound stores.

Figure 10.14 A solid-state digital sound store. The device illustrated can run three different sound tracks simultaneously, either on a continuous or demand-start basis. It requires no maintenance because there are no moving parts.

grumble that the maintenance of sound systems, especially after a few years' use, is an irritant. The cleaning and replacement of tape heads, the dubbing, loading and testing of cartridges, the frequent replacement of tape cassettes are all time-consuming and expensive, and, if not done properly, can drastically affect standards of presentation. 'Wouldn't it be nice,' they say, 'if sound could be stored in a box with no moving parts, requiring no maintenance and no consumables?' Now such a utopian device is available at a reasonable price.

As explained in more detail in Chapter 23, sound can be stored in either analog or digital form. The fact that it can be stored digitally allows it to be stored in standard memory integrated circuits (chips). However, the memory capacity needed is enormous, and is dependent on the frequency response required. Until recently the use of chips to store digital sound was both technically and economically impractical; technically, because to record a message of any significant length needed hundreds, if not thousands, of chips; and economically because such a memory would cost thousands of dollars.

Advances in technology now mean that one chip can store one million bits of information, and the cost per million bits is only a few dollars. This has allowed the introduction of digital sound stores which store the sound in non-volatile EPROM semiconductor memories. Such stores are entirely suitable for speech commentaries and sound effects, and even for music.

A distinction must be made here between speech synthesis and digital recording. Synthesis

may have a place in some special exhibits, but is of limited use to most users. The sound store should use a true recording technique that does not color the sound in any way. Typically, master tapes are made conventionally and then transferred to the chips using a special computer-controlled dubbing machine. The chips are then placed in the sound store, either individually or in blocks using a chip cartridge. From then on the sound can be run continuously for years with no loss in quality, and no danger of loss due to power failure etc.

Typical characteristics of a digital sound store, based on one particular production unit, are listed below. As the technology develops there will be many variants.

☐ Option of audio bandwidth of 3.8 kHz, 5.5 kHz or 11 kHz.

☐ Uses 256K or 512K EPROM chips for sound storage.

☐ Uses 6K bytes per second for 3.8 kHz bandwidth (compressed) or 15K bytes per second for 5.5 kHz (uncompressed).

☐ Typical capacity of standard unit between 48 and 600 seconds.

☐ Dynamic range 52 dB.

☐ One sound store can simultaneously process three low-bandwidth messages, so one unit can play three different sound effects, or one message in three languages.

☐ Instant start and restart, and continuous run or demand start.

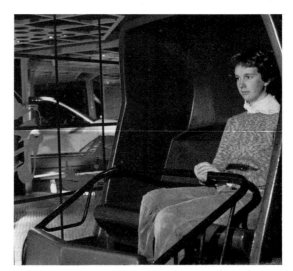

Figure 10.15 The 'Wheels' dark ride at the National Motor Museum, Beaulieu, England, uses a solid-state sound system. Each pod in the ride is fitted with its own commentary loudspeaker which keeps in exact synchronization with the exhibits being passed.

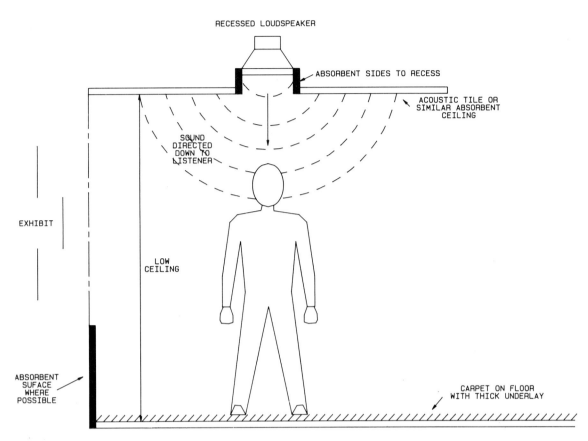

RECESSED LOUDSPEAKER

ABSORBENT SIDES TO RECESS

ACOUSTIC TILE OR
SIMILAR ABSORBENT
CEILING

SOUND
DIRECTED
DOWN TO
LISTENER

EXHIBIT

LOW
CEILING

ABSORBENT
SUFACE
WHERE
POSSIBLE

CARPET ON FLOOR
WITH THICK UNDERLAY

Figure 10.16 Ensuring that loudspeaker sound does not spill into adjacent areas depends entirely on appropriate sound absorption. This in turn means low ceilings, directional loudspeakers and, most of all, a carpeted floor.

☐ Possibility of selection of 64 message segments from three outputs.

The specification outlined above conceals an enormous range of applications. The frequency response of the basic unit may not seem very exciting, but it is similar to that of AM radio. The sound has a full bass response and is free of detectable noise. It is therefore quite suitable for the majority of commentary and effects work. (A tape cartridge will not give a significantly better response after a few months use!) Where a better response is required, it can be had at the expense of more memory.

Sound stores can hold other control data in addition to the sound, so they can also be used for running sound and light sequences. In the future it is likely that short multi-image shows will also

be run from sound stores. Therefore any new museum or permanent exhibition installation should seriously consider confining sound sources to either optical disks (videodisk, data disk or optical ROM, or compact audio disk) or to solid-state sound stores.

The acoustic guide

Sound can enhance a museum or exhibition, but it can also be a nuisance if it leaks to where it is not wanted.

Where there is a major show, or where there is ambient sound, loudspeakers are used for delivering the sound. The choice of loudspeaker should be made on the normal cost/performance criteria.

Figure 10.17 At the National Air and Space Museum in Washington DC, visitors can rent rugged compact cassette tape recorders to give them an acoustic guided tour.

The problem arises when there are a number of exhibits close together where one may interfere with another.

Loudspeakers still remain the preferred choice because they require no maintenance and no special action on the part of the visitor. Provided the exhibition is carpeted it is surprising how close it is possible to have exhibits with their own loudspeaker commentary, especially if care is taken to point loudspeakers down onto the listeners, and to use strategically placed display screens to isolate the sound.

Ultimately the problem remains, either because the museum management do not want a lot of commentaries on loudspeakers (e.g. in a picture gallery) or because of the need to offer a choice of languages. The choice then becomes that of either having earphones attached to the exhibit, or of giving visitors their own means of hearing the commentary.

One or more sets of earphones attached to the exhibit is technically the simplest, and also the cheapest, method. The phones will normally be similar to a telephone handset, or be of the lorgnette or stick variety. The obvious disadvantages are that they require the exhibit designer to allocate exhibit space to accommodate them, that the number of people listening to one exhibit commentary at one time is limited, and that the phones, especially their cables, are subject to wear and tear.

Another possibility is to fit the exhibits with an acoustic jack, and to issue each visitor with a pair of acoustic headphones of the type used on airliners. Such an approach is really only valid if a

charge is being made for the headphones, because it has no significant advantage over the attached phone method, and requires the extra work needed to issue and collect the headphones.

Issue and collection is a problem of any system which requires visitors to take devices found with them. It is at its worst when visitors are issued with small cassette tape recorders, a system that has been tried in several museums. The major disadvantages are:

☐ Serious wear and tear on the equipment, especially on cables.

☐ Headphones are not convenient for visitors.

☐ Visitors have to carefully follow operating instructions. If they do not, they get out of sync with the exhibits.

☐ It is not possible to run an audio commentary synchronized with an exhibit show (e.g. video).

☐ There must be a cash deposit system to ensure the return of the recorder.

The system with the fewest disadvantages, and, with new technology, actually good and effective, is a wireless transmission system. Here each visitor is given a small receiver, and wherever they stand in the museum they hear a commentary related to the exhibit nearest to them. The ideal attributes of this type of system are:

☐ The receiver has no controls and no loose wires for earphones.

Figure 10.18 At the Science Museum in London, England, handsets are used for exhibit commentaries. These commentaries should never be longer than two minutes.

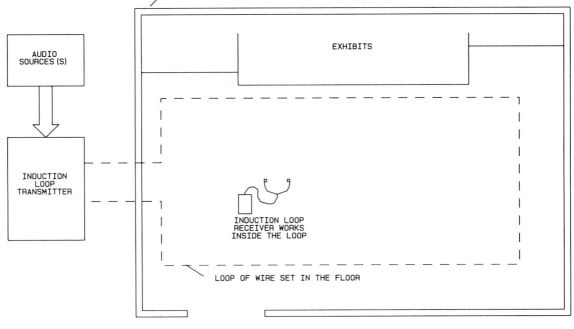

Figure 10.19 The principle of induction-loop wireless sound transmission. There can be a problem of interference between zones.

☐ The receiver uses rechargeable batteries with sufficient power for all-day use and overnight recharge.
☐ The sound quality can be made good enough for speech and even music, without any background noise.
☐ There is no crosstalk between adjacent zones.
☐ Where there is no sound, the system cuts out cleanly on leaving a sound zone, without any whistles, crackles etc.
☐ The system permits multi-language operation.

From the user's point of view the acoustic stick is the best device to give the visitor. This can have clean lines, and is easy to use. The visitor simply holds it to his ear, and differing hairstyles should be no problem. Although it is possible to have volume and language selector controls on the stick, it is better not to. Volume can be adjusted simply by how close the stick is held to the ear. Language selection is best achieved by having different colored receivers for each language, or by having a secret staff-operated selection switch. The less there is for the visitor to play with, the less likelihood of user difficulty and complaint. When not in use the sticks are stored in a combined storage and battery-charging rack.

The problem then is to decide on the best method of wireless transmission. Until fairly recently the only practical method was the induction loop. This has the major disadvantage that it is difficult to have several zones close together without mutual interference. The problem is especially difficult in an exhibition which is on several levels, because putting one induction loop above another one always results in interference.

The solution has arrived in the shape of infra-red transmission. The combination of low-sensitivity receivers and low-power infra-red transmitters allows very close spacing of sound zones. Modern receivers based on cellular radio technology ensure a clean cut into silence between zones. These installations must be carefully planned, but in principle, they can be extremely effective. The transmitters are usually overhead, projecting a cone of infra-red radiation. Anyone holding a receiver with upward-facing infra-red detectors and standing within the cone hears the message. Multiple transmitters are installed to cover the required area.

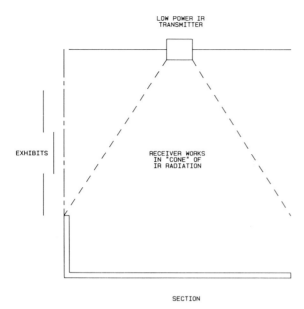
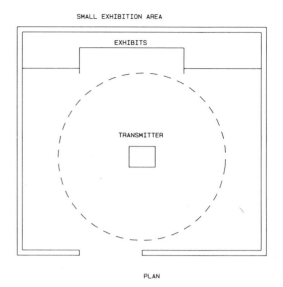

Figure 10.20 Infra-red wireless sound transmission has the advantage that it can be matched to a small exhibit area. This is because radiation behaves like light.

The combination of the solid-state sound store and the infra-red transmission system represents the current ultimate in acoustic guidance systems for museums.

Multi-media

At the beginning of this chapter brief reference was made to mixed-media shows. The possible combinations are infinite, but it is worth once again drawing attention to some changes in method that are taking place as a result of new technology.

It is easy to synchronize the action of several videodisk players (see Chapter 19) and equally easy to link these players to other media. Thus, a show that must link video images with multi-image slide and maybe also with controlled lighting could use a videotape as the master control medium. However, for a permanent installation it would be much better to use a disk. This in turn opens up new possibilities because for only a little extra money several separate video screens can be used. This can greatly help the exhibit design. For example, one screen may be carrying moving pictures, and another screen sub-titles or other support information.

For large-scale exhibits it is often best to think in terms of subsystems, each working independently but under common overall time control. The introduction of industry standard time codes has greatly simplified the architecture of big control systems such as those used in planetaria. The possibilities are discussed in more detail in Chapter 16.

Archive systems

Some museums have identified a need for study areas outside the main exhibit flow, where individuals or family groups can sit in a study carrel, and choose from a wide range of archival, program or interactive material. The material is usually presented on a video monitor.

Anyone embarking on the commissioning of such a system should not underestimate the enormous amount of work needed to make it fully effective. They should first consider the following questions:

☐ What will the system do that a good library and librarian would not do as well or better?

☐ Who is going to run the system?

☐ Who is going to keep it up to date?

☐ What exactly are the benefits to the visitor using it?

Provided the system is justified, it is probable that a disk-based system will be the most practical. Further information on this kind of system can be found in Chapter 12.

Show duration and audience circulation

When audio-visual shows are an integral part of an exhibit, it is essential to pay attention to their effect on visitor flow. Are *all* visitors expected to see a particular show? If not, there must be room for those who do not want to see the show to get around those who do. If yes, there must be sufficient viewing area to cope with the planned visitor flow.

For example, if shows are to be given in a small auditorium its size must be matched to visitor flow. In some cases this will be related to the way in which visitors arrive. Thus, the busload may be the unit which determines auditorium size. How long do visitors have to wait to see a show? Anything longer than ten minutes demands that the waiting area has some exhibit features of its own.

These kinds of questions are important not just for the successful running of AV, but also for the successful operation of any kind of visitors' center.

Chapter 11

The presentation room

A poor presentation environment creates frustration for the audience and presenter alike. Previous chapters have reviewed the use of AV in exhibitions and public venues; this chapter is devoted to the use of AV in the commercial field, and how to eliminate the problems of the presentation environment.

Frequently a commercial presentation must be given either in your own offices, or in those of a customer. If it is an individual presentation, it may literally have to be given in an office, but this should be avoided if possible. The very least that is required is a meeting room.

The serious presenter always:

☐ Has someone visit the room beforehand to check blackout, screen arrangements, seating, power points, etc.

☐ Arrives early to ensure a smooth set-up.

☐ Has a quick technical rehearsal before the audience arrives.

Any organization that takes presentations seriously soon recognizes the need for a *presentation room*. Ideally this is a room that is used solely for presentations, but obviously, in a smaller company, it may also do duty as a boardroom or general meeting room.

The presentation room creates the right atmosphere for effective communication. Correctly designed, it will make the presenter's job easier and the audience more receptive. The major benefit of the presentation room is *effective communication* with, for example:

☐ *customers*, resulting in more or sustained sales;

☐ *employees*, resulting in a more effective and better informed workforce;

☐ *suppliers*, resulting in better terms of supply and better service;

☐ *press*, resulting in better public relations and accurate reports of the business;

☐ *public* (if applicable), resulting in a well-disposed public and hence customers.

Presentation rooms can also be used to greatly improve the efficiency of management meetings. If meeting participants are introduced to the discipline of preparing all their material in a form that can be shown to a group in a presentation room, the other participants' time will be saved, and they will all be working from the same data. Often the only reason that this approach does not work is the reluctance of the participants to make the effort to prepare the material in the correct form. Nowadays there is really no excuse for this, because many visual aids, such as overhead transparencies, computer text displays and slides, can be prepared almost instantaneously. Provided the meeting is planned, the support material can be there too.

To achieve the intended benefits of the presentation room, there are several aspects of the room that need special attention. Usually the room will be in a normal office space with restrictions on ceiling height etc. However, even with such restrictions, it is quite possible to create an acceptable presentation environment. The following topics must be considered.

Seating

This must be comfortable and must have good sightlines to the screens. Usually it will be theater-style with movable chairs, but it may be necessary to allow for schoolroom-style as an alternative if the room is also used for training, or boardroom-style if the room is also used for management meetings. The room should be big enough for the largest regular audience expected, but not too big because it must feel comfortable. Some possible layouts are shown in Figures 11.1–11.5. Some useful guidelines include:

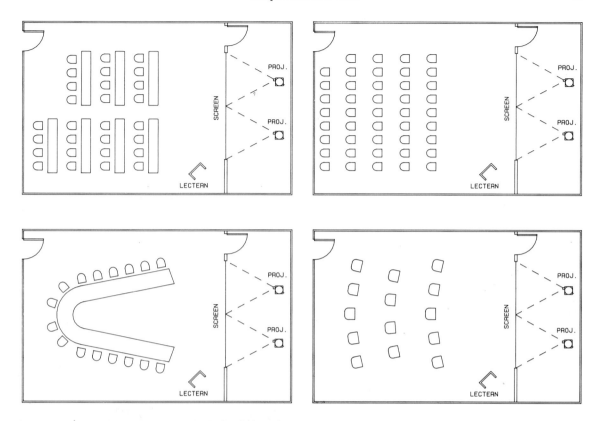

Figure 11.1 Four ways of using the same space. The nature of the intended meetings and presentations will determine whether schoolroom, theater, boardroom or island seating is appropriate.

Figure 11.2 Organizations giving regular presentations to large groups need a theater-style layout. (Photo of their Chiswick, England, presentation room courtesy IBM)

Figure 11.3 Some presentation rooms must provide writing facilities for the participants. (Photo of their Toronto customer presentation room courtesy Cantel Inc.)

Figure 11.4 When AV is used
to support meetings, the
boardroom layout is
appropriate. Notice the twin
screens for comparisons.
(Photo of their London,
England, boardroom courtesy
Phillips Petroleum)

Figure 11.5 Prestige selling
presentations to selected
corporate customers demand
island seating. (Photo of their
Toronto presentation room
courtesy Bell Canada)

□ A room about 7.5 m × 9.6 m (25 ft × 32 ft) will accommodate up to 45 people theater-style or 24 people schoolroom-style.

□ A room about 6 m × 7.5 m (20 ft × 25 ft) will accommodate up to 30 people theater-style or 15 people schoolroom-style.

□ For audiences up to 25 people the room should be at least 6 m (20 ft) wide.

□ For large audiences and in auditoria the practical minima for theater-style seating are 90 cm (3 ft) between rows, 54 cm (21 in) seat width and 106 cm ((3 ft 6 in aisle width.

□ The audience should not be outside a 60 degree viewing area (i.e. 30 degrees either side of center line). This may mean shorter seat rows at the front.

Lighting

The room must black out properly. For this reason presentation rooms are ideal for using up space in the building core or basement! All lighting must be under dimmer control, and the dimmers should be push-button automatic, *not* rotary knob or slider, because they will usually need to be operated from more than one place. The automatic dimmer ensures a smooth fade at all times.

In presentation rooms that are multi-purpose, or which use several different media, the lighting should be multi-scene. Usually this type of room will have several different light sources, e.g. downlighters, fluorescent, separate lights for

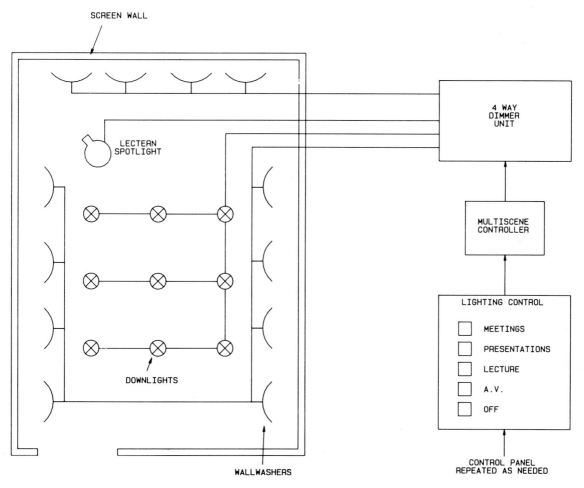

Figure 11.6 Block diagram of a multi-scene lighting control system for a small presentation room.

Figure 11.7 Push-button lighting control is essential for presentation rooms. It allows consistent results to be obtained from several different control points.

lighting the presenter, and display lights. It can then be confusing to have a separate set of controls for each. It is much better to define the various ways in which the room can be used and allocate one 'scene' for each use.

For example, there might be a scene for meetings, when a boardroom table is in use and no visual aids are required. This would be full lighting. Another scene might be for a lecture. In this case the main room lights would be dimmed, but would still be bright enough to take notes by, a lectern spotlight would be on, and all other lighting adjusted to ensure that no unwanted light fell on the screen. Other scenes could relate to video, product display, presentations and so on. The important thing is that there be only *one button* for each scene.

Lighting control will normally be effected from the lectern, from controls at the entrance door(s) and maybe from some central control panel. If there is a separate projection room it must be possible to control the lighting from there too.

Audio

The AV equipment will have its own sound system. In small presentation rooms it is often best to keep each item separate with its own loud-speaker. If there are many AV sound sources, or it is a big room, better results will be obtained by a specialist installation to mix the various sources.

The program sound must be completely separate from any speech reinforcement.

Because of the high absorption of sound by people and furnishings, even quite small presenta-tion rooms need speech reinforcement to reduce listener fatigue. This can be done simply, either by using ceiling loudspeakers, or by a lectern amplifier.

One of the main user problems in presentation rooms can be the complexity of the audio systems. Sometimes a consultant or supplier will, in good faith, recommend an audio system of the highest quality which the end user has difficulty operating because of its complexity. Very few users of presentation rooms can afford to have an operator present whenever the room is in use, so the ideal is clearly to have a system that operates itself.

This is now feasible. A typical prestige presenta-tion room would now be fitted with a speech reinforcement system with as many microphones as required, arranged so that the microphones are automatically controlled, i.e. if no-one is speaking into them, they automatically switch off. The system also ensures optimum microphone gain, and never goes into feedback or howlround.

The program sound system would usually be a high-quality stereo system playing through studio or auditorium-type loudspeakers placed correctly in relation to the screen. This system would also have an automatic mixer which is muted except when playing show material. It might well have to deal with several sources, such as tape recorder, multi-image sound, video sound and so on. It would also ensure that the speech reinforcement system was off while shows were being run.

A feature that is usually required is the remote control of show sound volume. This is best done with automatic push-button control because this

Figure 11.8 An audio mixer specially designed for the presentation room. It allows for automatic sound routing without the need for an operator.

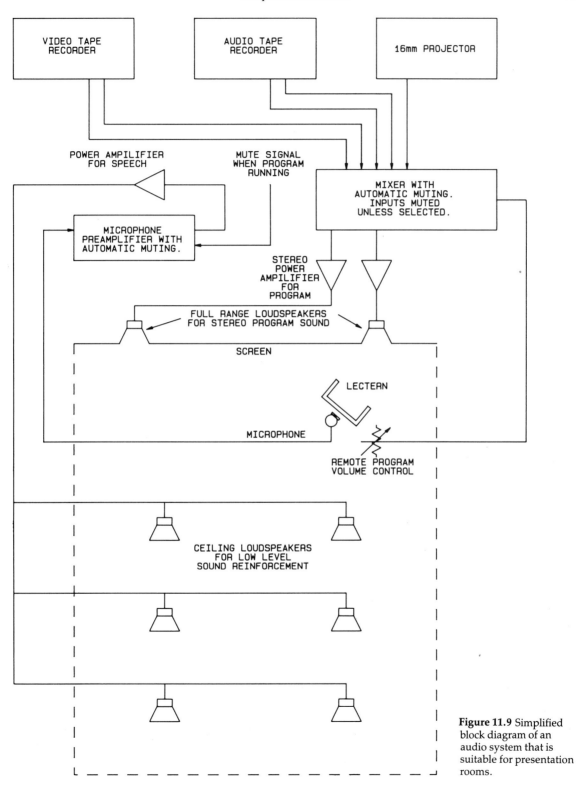

Figure 11.9 Simplified block diagram of an audio system that is suitable for presentation rooms.

kind of system can always be reset to a preset volume for the next show, thus avoiding the problem that a presenter might spoil a succeeding show by mistake.

The aim of good audio design in a presentation room is to make the user unaware of any technical complexity.

Lectern

All presentation rooms intended for formal presentations have to have a lectern. Many standard lecterns are now available, although often a user prefers to have one designed by his architect or interior designer to match the room. The lectern should give the presenter confidence, and act as a focus for the audience.

The lectern should include the following items:

☐ A reading light. However, this must be on the same dimmer circuit as the lectern spotlight to ensure that is goes out whenever an AV show is run. Otherwise it may throw unwanted light on the screen.

☐ Basic room lighting controls (i.e. the main 'scene' buttons).

Figure 11.10 For most presentation rooms it is advisable to keep controls on the lectern to a minimum. (Photo Multivision Electrosonic Ltd (Canada))

☐ Simple remote-control facilities for slides and for starting AV shows.

☐ A microphone.

The lectern can include many more items. The general rule is that if the lectern is going to be used by many different presenters, it should be kept as simple as possible. On the other hand if only one or two presenters are going to use it on a regular basis, they may appreciate more extensive facilities. Some of the many possibilities are:

☐ clock and/or digital timer;

☐ motorized height adjustment;

☐ control of motorized curtains and other display devices;

☐ video prompting system;

☐ video monitor;

☐ video writing facility;

☐ random access slide selection;

☐ random access video selection;

☐ push-button control of audio volume.

Many of these are discussed in more detail below.

Screens

Good presentation room design results in the audience being unaware of the technical equipment. They should be interested in the message, not its means of communication. Therefore, if at all possible, there should be a separate projection room.

If many different formats are being used, back projection may be best because:

☐ It allows projection to be used in relatively high ambient light conditions. This is essential for meetings.

☐ It is easier to mask to ensure that the screen is the right size for the medium being used.

☐ It allows a much bigger image where there are low ceilings.

☐ The audience cannot see any projection equipment.

In principle, a back-projection room should take up no more space than a front-projection room. Some other points to consider include:

☐ Whether using back or front projection, some form of curtains or cover panels should cover the screen(s) when not in use. Preferably, but not necessarily, motorized. Blank screens must be avoided.

☐ If using back projection, special lenses are available to limit the projection distance and to eliminate keystone distortion.

☐ Small presentation rooms can use a projection wall with built-in back projection AV devices to further

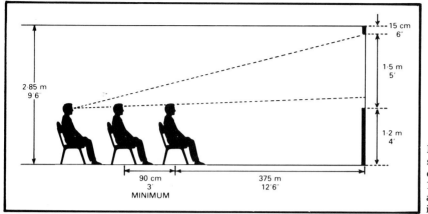

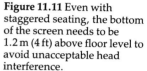

Figure 11.11 Even with staggered seating, the bottom of the screen needs to be 1.2 m (4 ft) above floor level to avoid unacceptable head interference.

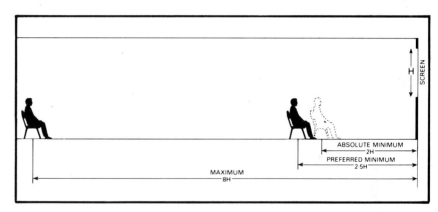

Figure 11.12 The audience should not be further away than eight times the picture height.

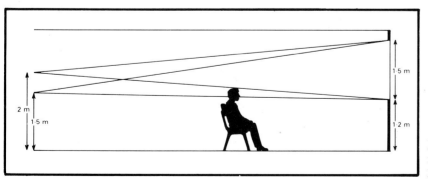

Figure 11.13 When front projecting, the projectors must be well up. The convenient operating height of 1.5 m (5 ft) is too low to avoid head interference. Back projection is often a preferred arrangement in presentation rooms to avoid this kind of problem.

reduce space. Alternatively, front projection equipment can be mounted in a suitable wall cupboard with projection ports.

☐ If using front projection from a projection room, the projection window should be full length for maximum flexibility.

The question of screen format naturally depends on the way in which the room is to be used. In practice, two screen areas side by side is a popular arrangement because it allows for the comparison of images – for example, two slide images, or one slide image and one video projected image.

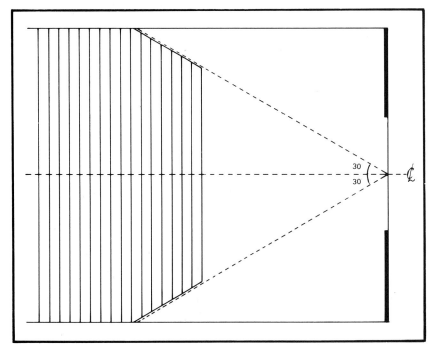

Figure 11.14 The audience should be within 30 degrees of the screen axis. In temporary installations with matt screens, 45 degrees is possible, but not recommended.

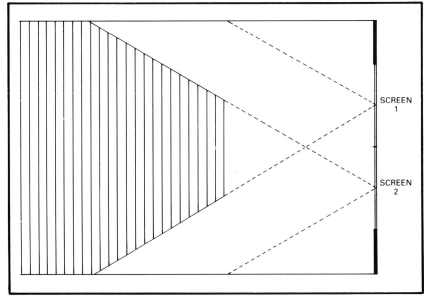

Figure 11.15 The 30 degree rule means that when there are two screens side by side the optimum viewing area is reduced.

Audio-visual equipment

Besides the lighting control, audio and screen equipment a presentation room will normally include at least:

☐ A single-slide projection system for visual aids work.

☐ A means of showing video programs, preferably projected, otherwise via monitor(s).

The problem is where to go from here. Provided the room has been designed with flexibility in mind there is no need to over-equip it at the outset, because extra items can be added when a proven need develops. The following observations give some guidance as to the possibilities, and recommendations as to the place within a presentation room of specific devices. The properties of individual subsystems (e.g. video presentation systems) are dealt with in more detail in other chapters.

Overhead projector

One way of distinguishing a presentation room from a training room is that the training room will always be equipped with an overhead projector, whereas a presentation room should never have one. The aim should be that all presentations are given to a uniformly high standard. This, in turn, means that all presentation material should be back-projected on slides.

In practice it may not be feasible to enforce such a rule. Notwithstanding the fact that slides can be made instantly, there will always be the board member who insists on bringing in his own hand-crafted OHP transparencies. However, the introduction of the OHP projector can cause an unfortunate degree of clutter in the presentation room design. If the use of the room can be defined so that the OHP is not needed, so much the better. If occasional use is unavoidable, it is important that a suitable screen arrangement is incorporated in the initial room design.

One very expensive possibility which has been used in boardrooms in the USA is to use a back-projection OHP. This device allows the presenter to face the audience and use the OHP in a conventional way. The projector itself is fitted with a very long focal length lens, which is directed at an opening to the side of the screen. Large mirrors are then used to get the image on to the back-projection screen. This arrangement is not only expensive in itself, but also makes the installation of slide and video projection equipment more difficult.

A more realistic alternative for many users is the video overhead. In principle, this is nothing more than a small video camera that is aimed at the document concerned; the resultant image is then presented by a video projector. In practice, the results are variable. In particular, it is difficult to get good and consistent results with actual OHP transparencies, which simply reinforces the point that it is better to prepare (or convert) the material as slides. However, if a properly designed top-lit system is used, quite acceptable results can be obtained from hard copy and even three-dimensional objects.

A typical system of this kind is fitted with a zoom lens and automatic focus, so that it can cover anything from an A3 document (about 30 cm × 42 cm, 11.5 in × 16.5 in) down to a small photo (about 8 cm × 10 cm, 3 in × 4 in). Such a device should never form the basis of a major presentation, but can be used in emergency if slides are not available, or for a spur of the moment item, and for presenting small products, packaging etc.

In the rare case of a user insisting that a video system be able to show standard OHP transparencies, it is best to use a system dedicated to this purpose. Such systems normally place the camera underneath the transparency, and rely on through lighting. The arrangement allows the presenter to write on the transparency in the normal way.

Another device is the video writer. This is an electronic writing board that allows the presenter to use a 'pen' on the face of a cathode-ray tube. The movements of the pen are converted directly to a video image. For most presentation room users, this type of device is really no more than an expensive toy, but it can have a place in rooms that are used for video conferencing.

Figure 11.16 The video overhead projector. When used with a video projector, this device can display conventional OHP transparencies using underlighting, or hard copy and three-dimensional objects using top or room lighting. It can also show slides.

Slide projection

Slides will form the basis of visual aids used in a presentation room. For many users, a single-slide projector is sufficient because this arrangement is the easiest to use and to prepare for. However, it is really much better to use a 'dissolve pair' of projectors, because this looks more professional and will properly exploit the build-up and sequence ability of slides. It should be no more difficult for the presenter to use, but does require the use of two slide trays.

All slides should be glass-mounted in the same type of slide mount. Autofocus systems and remote focus control should be avoided. However, there will always be some users who are unable to enforce this kind of discipline, and suffer from visiting presenters who arrive with a messy pile of card-mounted slides. If this is likely to be the case, remote focus control should be designed in from the outset as it is difficult to add later.

A popular format is the use of two slide screens side by side because this allows comparisons. In this case care should be taken to keep the presenter's remote control as simple as possible. It may also be necessary for the slide projection system be able to hold two or three projector shows, but unless these are permanently installed it is best to have an operator on hand to assist the presenter. This also applies if complex multi-image speaker support sequences are required, when the procedures become those of the sales conference.

An additional refinement is the use of random-access projectors. These allow push-button selection of any slide with an access time of about three seconds. In practice, the use of these devices in presentation rooms is limited because most presentations are linear. However, they are useful either in the meeting situation where a presenter is asked to recall a slide ('Let's see last year's figures again, Jim') or in the sales situation, where the course that a presentation takes is dependent on audience reaction.

The house show

Some presentation rooms can accommodate the house show. If an organization receives visitors on a regular basis, the running of an audio-visual presentation both saves time and gives the best introduction to the organization. Unless there is a separate visitors' center, the obvious place to run the house show is in the presentation room.

Because this type of show is unique to the organization, it may well be made as a multi-image show because this medium is the most flexible for updating and gives the best presentation quality. Even if it has been transferred to video for portable use, the show should be run in its original form in the presentation room. In order to ensure that it is easy to run, it should be installed on an auto-present basis (see Chapter 10).

A similar argument applies to movie; 16 mm movie, optically projected, gives a much brighter image, than the same material shown via a video projector. Thus, while video presentation will be used for all small rooms and where convenience of operation is required, any large presentation room, and particularly any organization which has commissioned material to be made in 16 mm, should show the film properly. The extra impact is well worthwhile.

Computer graphics and computer display

The role of computer graphics in creating visuals was discussed in Chapter 3. Here two problems specific to the presentation room are discussed.

It is a natural question to ask why, if visuals can be made on a computer, they cannot be kept in the computer and displayed directly from it. The answer is that they can, but the reasons why many users prefer to make a slide or overhead transparency for final use is that this way they get a better presentation quality and a much brighter image, usually at a lower cost. They also get something which is much easier to organize in to order.

Thus, anyone intending to use direct computer output needs to consider:

☐ Will the final image (from video projector) be bright enough, big enough and steady enough?

☐ How will the images be stored (hard disk, floppy disk etc.)?

☐ What computer program will be used to sort out the images so they come up in order like a slide show? Will this program allow last-minute changes, easy editing etc.?

The results that can at present be obtained by 'personal' computers are rather limited, but are certainly usable in the smaller presentation room. The main weakness is a lack of standardization and the practical difficulty, for anyone other than a regular user of the particular computer, of compiling a show.

For many users, the hybrid solution of creating a visual on a computer and then transferring this to slide, especially instant slide, is a very satisfactory one. Instant slide making kits are one of the tools of the trade for any well equipped presentation room.

A technical problem arises when a computer is connected to a video projector. Few computers have a standard video output like that used by a TV set. If a video projector is to be used for showing computer output, this must be specified. Most data projectors can also show a TV picture, but this is not the case when extra high resolution is required.

Such a requirement arises when the normal screen output of a computer must be displayed (e.g. a table of figures as seen by the computer operator, as opposed to a proper visual). On the whole, this practice should be discouraged, because, for all the reasons discussed in Chapter 2, nobody should use raw data in a presentation; they should create a proper visual. However, it can be legitimately argued that a weekly board meeting, arranged so that everyone is near enough to the screen to be able to read the figures, might require access to the company mainframe computer. If this is the case, only projectors intended for this application should be used.

Room control

What starts out as a simple concept, a room to give presentations in, can develop into something of a monster. The apparently simple requirement for the audience to be able to see and hear everything properly; for the presenter to be able to show a series of visuals; and for it to be possible, for example, to be able to run a movie, show a videotape, see a program on TV or display a teletext page can result in a control panel that looks like the cockpit of a 747.

The room itself should be designed so that the audience are not aware of specific technologies, only of the information that is being communicated. However, users can be all too well aware, especially if, for example, they are trying to juggle a lectern control panel, a slide projector control, and three different infra-red controls for different parts of the video system. To minimize this problem, the correct procedure is to have a room control system that minimizes the number of controls presented to the user.

An example that was mentioned earlier is the idea of 'scenes' for the lighting. In a larger or more sophisticated room the principle can be extended to cover the operation of the entire room. Modern presentation rooms use room controllers that are programmed to meet the needs of the individual user.

The principle is that each device in the room is able to operate independently (this is essential to allow for maintenance and the, happily remote, possibility of total system failure) but is also linked to the room controller. This is a microprocessor device that can be expanded to work with any foreseeable combination of equipment. The devices could be:

☐ lighting dimmers;
☐ motorized panels;
☐ blackout blinds.
☐ curtains;
☐ turntables;
☐ video projectors;
☐ video monitors;
☐ videotape and disk players;
☐ display computers;
☐ slide projectors;
☐ movie projectors;
☐ anything that can be electrically controlled.

The running of a videotape may require the following sequence:

☐ Switch on the video projector and videotape player (if not already on).

Figure 11.17 Infra-red wireless hand controls are a more convenient way of controlling presentation room facilities. When used with a suitable room controller the number of buttons can be kept to a minimum.

☐ Close the blackout blinds.

☐ Fade the lighting to video scene.

☐ Open the screen curtains.

☐ Start the videotape.

It would obviously be much easier for the presenter if he only has to press one button labelled start videotape, than it would be if he had to remember all the other buttons that had to be pressed. This is the facility that the room controller can provide.

This kind of approach makes it more likely that the room will be used properly. It greatly reduces the number of user controls, and makes use of the room less intimidating. It can include all the necessary logic to ensure that everything is in the correct state for a particular part of a presentation. Provided the system is of the programmable kind, it allows for changes in the way the room works if, in practice, it is found that different timings or level settings are needed.

One system is delivered as a compact rack-mounting unit for permanent installation. Programming is done on a personal computer; in this way different logic rules, settings and timings can be tried out. The room is first modelled on the computer, so it is often possible for the user to spot snags before the main system is built. Once programming is complete, the program is transferred to the room controller in the form of non-volatile PROM memory.

Remote control

An incidental advantage of the room controller concept is that it can greatly simplify control wiring. Typically, a lectern control panel of any complexity needs only four thin wires. This allows the room to be more flexibly laid out, with possible control points spread round the room.

A further possibility, not confined to rooms with a room controller, is to use infra-red wireless remote controls. These are available in forms specially suited to presentation use. They have the following characteristics:

☐ Typical controls have a 30 m (100 ft) range.

☐ They are specially engraved to match the requirements of the particular installation.

☐ Hand control versions control about fifteen different functions.

☐ Desktop or lap controllers control 30 or 45 functions.

☐ When used in conjunction with a room controller the number of functions can often be kept down to fifteen or less.

Conclusion

A presentation room is more than the sum of its parts. It is essential to have the technical aspects of a presentation room engineered as one system, so it is usually best to make one supplier responsible for its successful operation.

Chapter 12

Interactive audio-visual

Interactive audio-visual is fashionable. People who have not used it or who do not know what it is are worried that they may be missing out on something. It is true that some kinds of interactive AV are leading to new ideas in point-of-sale and other display applications. It is also true that in the training area, where interactive AV is likely to be of the greatest benefit, it is very important that any intending practitioner is fully conversant with linear AV methods first.

The essential differences between normal AV and interactive AV are:

□ A normal AV program is linear, it has a beginning, middle and end and is seen in its entirety. Interactive AV is non-linear, in that a given presentation is unlikely to use all the material available, and the order of presentation can be different for every viewer.

□ A normal AV program does not require any input from the viewer. The viewer is passive. Interactive AV requires the active participation of the viewer.

□ The participation itself can be on two levels. At the basic level, the viewer is simply making choices, using the random-access facility which is at the heart of any interactive system. More truly interactive systems require the viewer to give information about himself.

Generally, interactive AV is intended for an audience of one. The resulting presentation matches the one particular viewer's requirement. It is possible to run a large audience show, with branching points in the show determined by a majority vote of the audience, but such programs are necessarily fairly simple and certainly do not represent the mainstream use of interactive AV.

Interactive AV depends on the random access of program material. In response to the direct request of the viewer, or as a pre-programmed requirement based on an evaluation of the viewer, it must be possible to retrieve segments of material. These can be still images, moving images, or audio segments. Branching within a program arises when a program reaches a stop point. There may

then be several different ways forward. The choice of which branch to take can depend on the choice of the viewer, but in training applications it is more likely to depend on an assessment of the viewer's knowledge. A typical situation might be where the trainee is asked a question. If he gets the answer right, the program continues with the next linear segment; if wrong, the program loops, to repeat the previous segment. If he gets the answer wrong again, the program branches to a remedial program, only coming back to the main line when this has been successfully completed.

These programs are not easy to write, and it can be important to monitor how each presentation is actually run. If trainees have consistent difficulty with a particular question it may turn out that there is a deficiency in the program rather than in trainee!

Flowcharts

Any interactive AV system consists of a minimum of four components:

□ a visual display;

□ a store of images;

□ an input device;

□ a device to determine what is to happen as a result of the input stimuli.

Strictly speaking, if it is an AV system (as opposed to an interactive visual display) there is also a store of sound segments. Finally, there can be additional output devices. Thus, one particular configuration could have a video monitor as the display device; a videotape with a random-access search system as the sound and image store; a personal computer as the system organizer; a keyboard as the viewer input; and a printer as the additional output. This is only one of many possibilities. Another system could be built up using an entirely different set of devices. However, before letting the hardware impose any limitations, it is instructive to examine the nature of an interactive program in more detail.

As with all AV programs the intending user must be quite specific about what the viewer is going to gain by participating in the program, and how quickly it is reasonable for this objective to be met. For example, the following types of programs might be developed:

☐ A program designed to teach people how to complete computer records in inter-company transactions.

☐ A program to encourage shoppers to buy from a range of garden products.

☐ A program to allow museum visitors to find out what is in the reserve collection.

☐ A program to train an automobile mechanic in the servicing of a particular model of car.

☐ A program to teach elements of microbiology.

These all represent perfectly valid requirements for interactive AV. The nature of the programs would vary greatly. The microbiology program would assume that the student had a reasonable amount of time to work through the course (hours or probably days) and would include many tests of the student's understanding. The public programs, like the point-of-sale program or the museum program, would have to work very quickly to be of value. Their success would be partly measured in how many people per hour could use the system, as well as in the increase in sales or visitor satisfaction.

It is convenient to describe a particular interactive program by a flowchart. This is in addition to the story-boarding of the AV segments. The flowchart shows how the program is intended to work. The flowchart should show exactly how the user gets in to the program, and, just as important, how he gets out of it. It has been known for producers to inadvertently introduce dead ends into a program, resulting in a complete system lock-up and a very confused viewer.

The process of flowcharting is best understood by reference to a simple example. The example can then be followed through to embrace greater complexity, and with the same example it is possible to see how a hardware choice is made.

The 'London Experience' is a multi-media show at Piccadilly Circus in London, England, intended for tourists. Shows start every 40 minutes, so there is a problem of entertaining those waiting to see a show. An interactive game about London was seen as something that would both amuse and educate visitors. Before any decisions were taken about the nature of the game, some immediate priorities were considered:

☐ It must be easy for the visitor to join in, and be quite obvious how it works.

☐ It must be possible for the game to be enjoyed and give information in both short and long sessions. A visitor might play it for thirty seconds or ten minutes.

☐ The response of the system to visitor input must be very fast.

☐ Assuming that (as is the case) three separate systems are put in, any observer should not see the same material repeated often within a waiting session. Equally, there is no point in having more program material than is necessary to meet the objective.

What could form the basis of the game? Something based on London sights obviously commends itself, so the 'London Game' is a simple quiz game with detailed answers to help visitors learn something about London.

Figure 12.1 shows the game flow chart:

☐ The rest state shows a text image telling the visitors what to do. In this case, they must press a start button.

☐ Pressing the start button brings up a choice of three subjects. The visitor chooses by pressing one of three buttons. (The choice is deliberately limited to ensure a fast response.)

☐ An image appears relating to the chosen subject. For example if the subject chosen was London Squares, an image of Berkeley Square might appear.

☐ Then, next to the image appears a multiple-choice question with three possible answers. The visitors express a choice by pressing the button corresponding to what they believe to be the right answer.

☐ If the answer is correct, text appears confirming that this is so and a score indicator shows a score of one correct answer.

☐ If the answer is wrong, text appears saying why, and giving the correct answer. A wrong score indication is shown.

☐ After a pause to allow for the text to be read, the procedure is repeated with a new question.

☐ At the end of five questions, a pause image appears, inviting the visitor to press the start button again if he wishes to continue.

Various housekeeping additions have to be made to the flow chart to deal with the following questions:

☐ How does the game finish?

☐ How to make sure there is sufficient variety of material?

☐ What happens if the visitor fails to respond?

In this case there need be no set finish to the game, the idea is that the visitor can go on playing

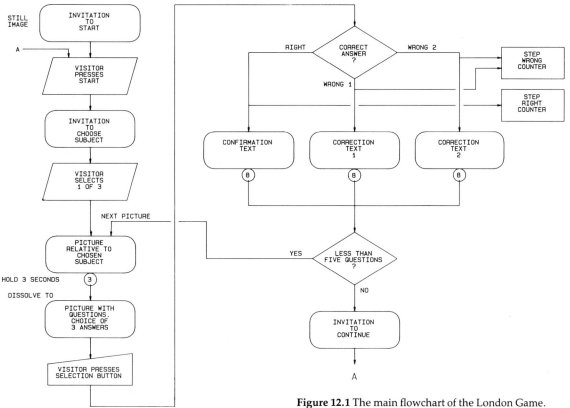

Figure 12.1 The main flowchart of the London Game.

until the next performance of the main show starts, or until he loses interest. In fact, the game does stop if one of the score counters reaches 99!

Although there is an initial choice of three subjects, this group of three is, in fact, one of six such groups, making eighteen groups of five questions in all. Each time the start button is pressed the system presents a group at random. This ensures that the average visitor is unlikely to see the same set of questions twice. If the visitor fails to respond within a set time delay, the score counters go back to zero, and the rest state image reappears.

The game meets the need at the London Experience. It is easy to embellish the game, although in that particular location there was no need to. The embellishments that could be added include:

☐ A prize being given for reaching a particular score.

☐ A printed card being issued on request that gives directions to get to a tourist site.

☐ A speeding-up of the presentation of the questions as the score increases.

The emphasis in the London Game is speed of response, and giving information in an interesting way. There is no *need* from the owner's point of view to find out whether the visitor has retained the information given to him. The flowchart for a training application can look completely different.

What can you see here:

A. Music Hall?

B. Experimental Theatre?

C. Restoration Comedy?

Figure 12.2 A typical question in the London Game.

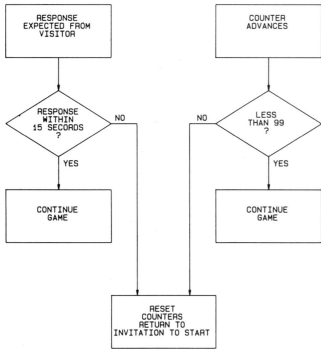

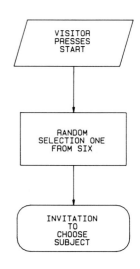

Figure 12.3 Management sub-routines in the London Game.

A USA automobile manufacturer wishes to train aspiring automobile mechanics in the servicing of car electrical equipment. Here, there is no question of needing to achieve results within the first few seconds. The interactive program is designed to be used over several hours of study, and, furthermore, it is to test the knowledge of the student and be the arbiter of whether he has passed an examination in automobile electrics.

A possible flowchart is shown in Figure 12.6. It has been greatly simplified, omitting most of the AV detail, but showing some of the major management differences. It is also broken down into sections to make it easier to understand what is happening. The following points distinguish this type of program:

☐ The system has to keep track of a student's progress. The first time that a student uses the program he must log on. His name can then only be removed from the log of current students by an authorized assessor. Equally, it would be possible to prevent anyone logging on in the first place unless their name was authorized.

☐ The first time the student logs on, a choice of language of instruction can be made. In the example, the choice is between English and Spanish. From now on all

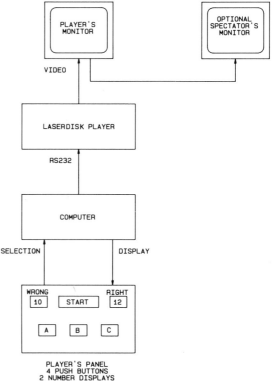

Figure 12.4 Block diagram of the hardware system used in the London Game.

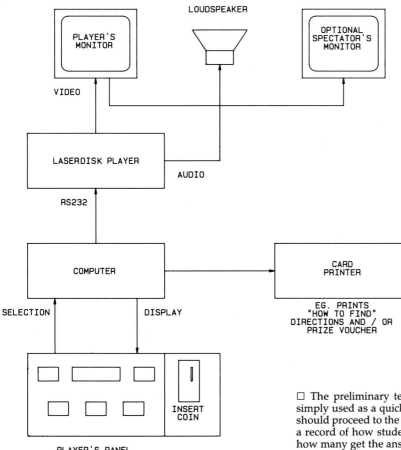

Figure 12.5 Possible embellishments to public information interactive displays like the London Game.

audio commentary and text graphics will be in the chosen language.

☐ The first item in the program is a general explanation of how the program works, and how the student should use the keyboard to respond to questions.

☐ Obviously, for the first-time student the program starts at the beginning. But, if the student has already had one session, the control device (nowadays invariably a computer) starts at the point previously left off. This part of the program needs designing with care to cater for those students who leave off studying in unexpected places. The technique is always to restart at least one segment earlier than recorded as complete.

☐ Most testing of the student is done by multiple-choice questions. Failure to get an answer right usually involves the repeat of one or more previous segments, and sometimes involves remedial loops.

☐ The preliminary testing is part of the course and is simply used as a quick check to see whether the student should proceed to the next segment. An option is to keep a record of how students respond to the questions (e.g. how many get the answers right first time). At this stage this procedure has less to do with assessing the student than with assessing the effectiveness of the program.

☐ An entirely separate exercise is the assessing of the student. The example shown is fairly sophisticated because an attempt is being made to ensure that the student really understands, and is not just learning by rote. At set points in the student's progress a series of questions is presented. Importantly, these are retrospective over the *whole* course so far, and they are random-accessed from a library of questions. This ensures that if a student has to resit the test the questions will not be identical.

☐ The student's responses are recorded for separate assessment if required. It is then optional as to whether the student is allowed access to the next part of the course if he has a satisfactory score, or whether such access is only permitted as a result of an external assessor confirming a pass. If the questions are only multiple choice, there is no reason why the test should not be fully automatic. However, if the test requires the student to answer some questions by text statements, the external assessment is likely to be essential unless the statements are extremely simple (in which case there is little point in using them).

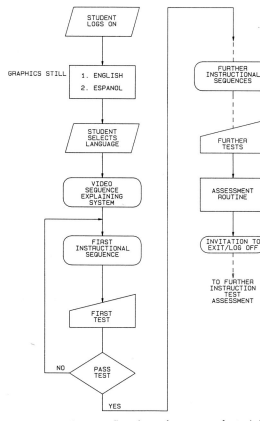

Figure 12.6 The main flowchart of a program for training automobile mechanics.

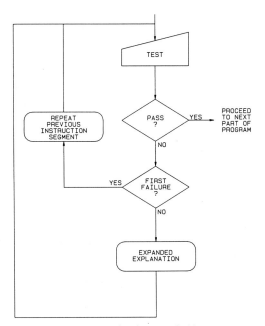

Figure 12.7 An example of a remedial loop.

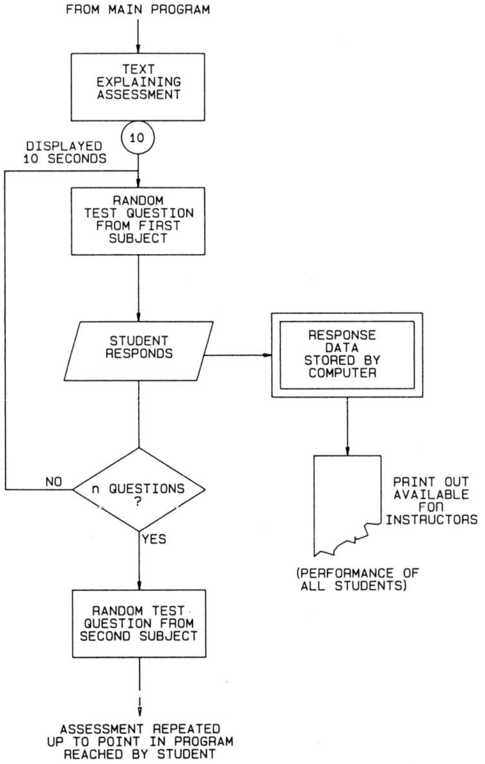

□ There must still be a device to cater for lack of response. Normally a student will be expected to log off at the end of a session. This is done with a single key stroke that simply records his progress so far and returns the system to its rest state. A similar sequence of events takes place if, at any time, there is no response from the student within a preset time.

The two examples given are almost two extremes; together they illustrate the possibilities of interactive AV and the importance of flowcharting how an intended program is supposed to achieve its objective. It is immediately clear that training and assessment programs can get extremely complicated, so first-time users are well advised to stick to simple objectives. This usually means breaking the task down into small segments, and getting these right one at a time. This may also reveal that some aspects are better dealt with by the separate presentation of linear programs.

Figure 12.8 An example of an assessment routine.

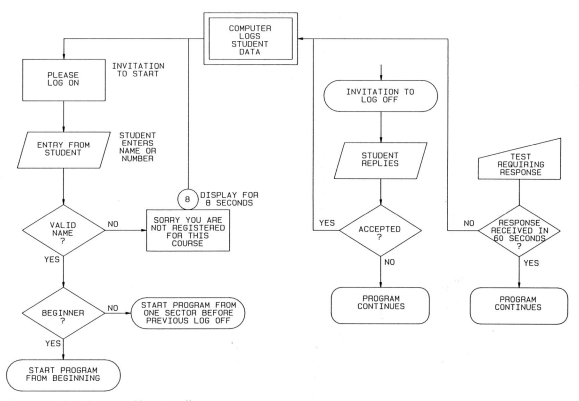

Figure 12.9 Logging on and logging off.

The hardware

Much current literature, whether it be journal articles, press reports or advertising, gives the impression that all interactive audio visual is video based. However, as Tables 12.1 and 12.2 show, there is a wide range of devices that can be incorporated. For reasons which will become clear, many applications now benefit from the use of videodisk, especially those requiring multiple users. However, there are many stand-alone applications, and even some multiple-user applications, that are better served by different or hybrid technologies. These are best explained by example.

Besides deciding on the best method of *delivery* of the interactive program, the intending user has to seriously consider the method of program *creation*. For most applications the cost of delivery systems can be seen as reasonable; very often it is the perceived cost of program creation that is the bar to proceedings. Before giving some examples that show the way in which a decision can be taken, the performance of some of the available system components will be discussed. A summary of interactive video sources and hardware is shown in Tables 12.1 and 12.2.

Random access slide projectors are convenient for the development of programs using a limited number of photographic visuals. They allow easy editing, but limited capacity (one unit for 80 slides). The random-access performance is reasonable, typically three seconds access time.

Videotape cassettes are convenient for program development when the program relies on moving images. The big disadvantages of videotape are the slow random-access performance (many seconds, or even minutes) and poor still-image performance. Thus, if still images are going to consist mainly of text, a solution to the problem is to use the videotape for moving images, and computer-generated text for the stills.

Random-access audio devices are available. The most accessible is the random-access cassette recorder, but as with the videotape unit access is slow. Digital audio stores based on hard disk

Table 12.1 Interactive audio-visual sources

Still images	Photographs
	Artwork
	Text
	Charts
	Computer text
	Computer graphics
	Freeze frame from moving picture sequence
Moving images	Movie film
	Animated cartoon
	Animated computer graphics
Other visual input	Combinations of the above, e.g.
	Text superimposed on movie
	Illuminated displays
Audio	Speech recordings
	Music recordings
	Sound effects recordings
	Speech synthesis
	Music synthesis
	Effects synthesis
Other output	Printed text
	Photographic output
	Plotted output

Table 12.2 Interactive audio-visual hardware

Visual display	Cathode-ray tube (CRT)
	Rear-projection screen
	Front-projection screen
	Illuminated display panel
	Liquid crystal panel
	Plasma panel
Visual source	Slide projector
	Motion-picture projector
	Videotape
	Magnetic video store
	Videodisk
	Computer text store
	Computer graphics store
Audio	Combined with visual source
	Cassette
	Synthesizer
	Digital audio store
Participant input	Custom control panel
	Standard keyboard
	Touch screen
	Mouse
	Light pen
	Speech recognition
	Credit-card reader
Other outputs	Paper printer
	Card printer
	Graphics plotter
Control	Dedicated controller
	General purpose computer
	Inherent in the display device

stores are the immediate answer; probably to be superseded by optical disk stores in the near future.

Videodisks seem to be the ideal basis of many interactive applications. Laser disks, in particular, have good still-frame performance. They can easily carry a mixture of moving and still material, have a large capacity (up to 54,000 still frames), have fast random access (sometimes instantaneous, usually between one and five seconds) and can carry a lot of audio information and even additional computer data. Their disadvantage is

the need to reduce all material to video, and the relatively high cost of having single disks made. Now most video disks are made in specialist plants from master tapes supplied by the user. Recordable disks that allow the user to record his own disks directly are becoming available, but are expensive.

Personal computers often form the basis of interactive AV systems. This means that other devices in the system must be of a kind that can be communicated with by a computer. In general this means that they must have an RS232 interface, which is a control connection able to receive computer data to an internationally agreed format. Random-access slide projectors, cassette recorders and videodisk players are all available with this as a standard feature. Videocassette machines usually need a special additional controller. However, the use of a general-purpose computer is not a *necessary* requirement of an interactive AV system. Often the final delivery system uses either a

Figure 12.10 A random-access cassette recorder suitable for computer control. (Photo Tandberg Ltd)

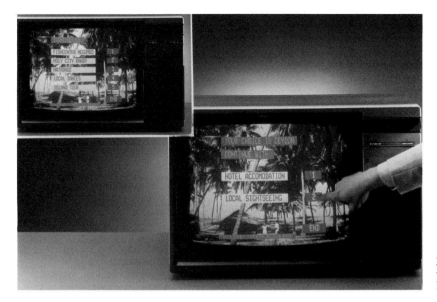

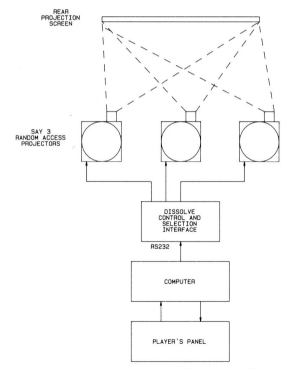

Figure 12.11 A touch screen in use. (Photo Philips Laservision)

special controller or a microprocessor controller within a videodisk player to achieve control.

Touch screens allow participants to express responses by touching the viewing screen at a designated point (as opposed to using a standard keyboard or special push-button set). They are usually used over CRT displays, but are equally applicable to flat-screen displays. They work in various ways. Some use a transparent overlay, which includes a grid of printed conductors. Changes in electrical resistance or capacitance locate where the touch has been made. Others use an array of infra-red emitters and detectors, which form an invisible X–Y grid across the face of the display. They are helpful in public displays and in any application where it is necessary to minimize the amount of hardware that can be touched by the participant. They make the preparation of the program slightly more complicated.

Examples of hardware configurations

In order to understand how the hardware configuration is conceived, first consider the London Game application cited earlier.

In its basic form, this game could have been based on two random-access slide projectors and a special controller. The controller could either be specially made, or consist of a standard computer able to control the projectors. There must also be a visitor's control panel fitted with a start button and three choice buttons. This method is recom-

Figure 12.12 Although the London Game was finally committed to videodisk, a display of the same type required in only one venue could be more economically, and in some cases more effectively, achieved with the equipment configuration shown here.

mended for anyone who requires a single installation, giving the highest picture quality, which may possibly need updating.

In fact The London Experience management elected to use videodisk. The reasons for this choice were:

☐ There was not space to install a slide-based system. Instead, the participant's panel was fitted with a small monitor, and another larger monitor was placed higher up for other visitors to see.

☐ The slide system would not have had enough capacity. Actually, at least six random-access projectors would have been needed, which while justifiable in a single installation, was uneconomic for the three complete systems needed.

☐ The random-access slide projection system would have needed more maintenance than the videodisk system and would have had a slightly slower response. Both these points were, however, only marginal considerations.

☐ It was possible to use the videodisk being prepared for the London Game for other purposes. The game only occupied part of the disk, the rest of it could be used for other material, thus lowering the cost of the disk attributable to the game.

The decision in this case was a marginal one. A museum wanting to create its own specialist display, and not requiring such a big image capacity, could well choose the optical projection alternative. In fact, the whole London Game was first made on slides, because this represented the most economical method of production. When the production was costed on a pure video production basis it was so expensive that the project would have had to be abandoned if pure video methods had been used. This simply demonstrates that the potential interactive user must be aware that there may be many ways in which the objective can be met, but only one or two of them will be practical or affordable.

Once the decision had been taken to use videodisk, the London Game *could* have used a touch screen. This was dismissed on the grounds of unnecessary extra cost, unnecessary extra programming, the fact that with only four buttons the input requirements were so simple, and finally because a display with some big buttons to thump was actually more appealing to the visitor than one with just a fingermarked screen.

The automobile electrics training program is the kind of program that needed the arrival of the videodisk as a mature medium to work at all. This

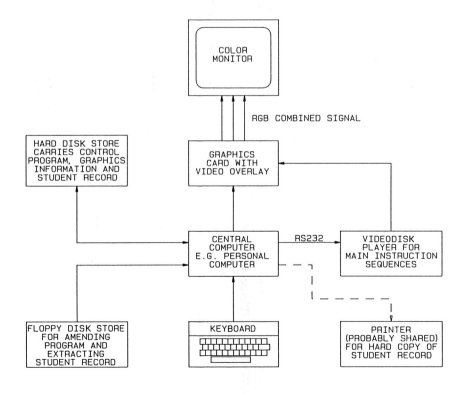

Figure 12.13 A possible configuration for the automobile electrics training program.

type of program would inevitably use both moving pictures (in the form of animated diagrams and film demonstrating procedures) and still images (as diagrams, photographs and text). A possible complete configuration is shown in Figure 12.13.

The computer, as opposed to a small processor in the videodisk player, is needed for this application because it is necessary to keep track of the students' performance. The computer keyboard is most likely to be used as the means of student response because this allows students to enter their names, and to answer questions in text format for subsequent retrieval by the assessor. A printer is likely to form part of the system, to extract student performance records in a printed form. This configuration is also ideal where a disk has been prepared that can be used in several different ways, because it allows the control software to be changed very easily from the computer store. If additional text, not on the videodisk, is required it can be created as a computer text overlay.

This last refinement is not available on all computer/videodisk combinations. The player must have the facility for text overlay, and the computer must be able to 'genlock' its graphic output, in order to be in sync with the disk picture.

Not all training applications are necessarily best served by videodisk. The teaching of in-house procedures for transaction processing, job progressing, capital equipment movements, etc. may be better dealt with by a computer-based system. The UK company Mast Learning Systems specializes in preparing customized programs for small users using the configuration shown in Figure 12.14.

A typical application could be a civil engineering equipment rental company. They have equipment rented out at many sites, and themselves have several depots which store the equipment not rented. They have a major problem in knowing where all their equipment is at any time, and whether it is in full working order or needs servicing. The problem is compounded by the fact that any mistakes made by their staff in following the company's documentation and data logging procedures, result in them losing track of individual items.

Therefore, all their staff have to learn the procedures. Because many of these involve entering information on a standard keyboard, and because they are all based on the use of either printed forms or data displayed on a VDU, it makes sense to use a personal computer as the

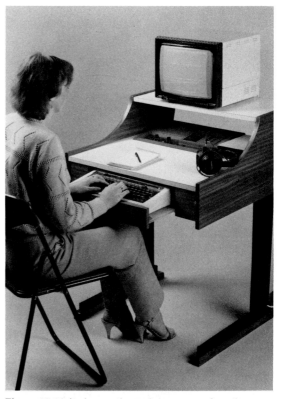

Figure 12.14 An interactive training system based on a computer and random access cassette recorder. (Photo Tandberg Ltd)

central training device. The program is presented in the form of text and computer graphics. The student responds by entering text on the keyboard, thereby simulating what they will be doing in real life later on, or by answering multiple-choice questions.

The computer alone would be rather dull, and would not give the student the vital sense of participation that is needed to make interactive AV a success. This is why the system also includes a random-access cassette recorder. This allows an instructor to take the student through the material with a voice commentary and, where a form is being described on the computer screen, an arrow or other attention-getting device can follow the commentary. Similarly, the instructor can ask the questions, and comment on the student's response, verbally, which is much more friendly. The random-access cassette recorder is sufficient for simple programs, but for those with many sound segments a magnetic or optical disk store would be better.

A feature of programs such as these is that they are highly specific, with a very small audience. They must be easy to modify, and their cost of preparation must be reasonable. The videodisk would be too expensive and too inflexible for this type of application, but that is no reason why such users cannot have the benefit of interactive training. Some of these programs will also need still images, and while some still images could be generated in the form of a computer graphic, most users would find it quicker and easier to use a slide. It is a simple matter for the computer to control a self-contained random-access slide unit, and these programs should never need more slides than can be held in a single magazine. If they do, it should be back to the videodisk!

Authoring

This last example has exposed the Achilles heel of interactive AV. It is all very well to describe how a system is intended to work, but how does an individual user, even one experienced in conventional AV production, actually put a program together?

For the major corporation requiring a large

Figure 12.15 Teletape Video is one of the companies offering complete systems for interactive video. Their MIC system includes a special add-on card for the IBM PC and compatibles that makes the computer program largely independent of the precise hardware configuration used.

number of copies of an interactive program on videodisk, there is no great problem. They can define their needs and go to a specialist producer who will do the whole job for them, and be happy to relieve them of the $100,000 or so that it may cost. For the small user this approach is out of the question, although some specialist help may be necessary to achieve cost-effective results.

The main problem is the need to have a control program in addition to the AV software. The control program may be machine code software buried in a microprocessor, or, more likely, a program in a higher level language on a personal computer. But, even in this case, it should not be necessary for an intending user to have to learn BASIC or PASCAL in order to get started.

There are a number of authoring systems available that allow users to get straight down to making interactive programs. Some are dedicated to particular hardware packages (for example Sony have a complete package system), others are more general. For example, there are several available to run on Apple IIE computers and on IBM PC personal computers. The IBM programs usually allow a choice of ancillary equipment.

The authoring system itself is nothing more than a computer program that prompts the user to feed in the parameters of his presentation. Flowcharting must still be done, but a good authoring system will prevent anything being left out or the creation of dead ends. It also allows the setting up of computer text to augment the video presentation, and allows the setting up of student tests.

Interactive systems based on standard equipment with an authoring package represent a good starting point for the user who expects to develop applications in-house. For those requiring delivery systems only there may be other more cost-effective solutions. Systems embodying personal computers make sense in those applications where a lot of information must be extracted from the participant, or where programs are frequently changed, but many display applications can be better served by simpler equipment.

An example of equipment packaging that, for the ultimate user of an interactive AV, does away with the need for a computer (even though programs for it are authored on a computer like the IBM PC) is the Philips VP415 Professional Laserversion player. It has a microprocessor controller built in, and has a facility for specially created control software to be either dumped into it from the videodisk itself, or inserted by the use of a PROM cartridge. The unit can be used with a touch screen, and allows for text overlay.

The Nebraska scale

The Nebraska Videodisk Design and Production Group were one of the pioneers of interactive video. They devised a scale of interactivity which is widely referred to. It can apply equally well to videotape-based or videodisk-based systems.

☐ *Level 0* refers to equipment that has no real interactive capability at all. This would apply to a domestic videotape recorder, or some domestic videodisk players.

☐ *Level 1* refers to equipment that has basic random-access capabilities, controlled from the device's own control panel. It would be expected to include at least random access of any required still frame and be able to play any specified program segment from one designated frame to another, at either full speed or a designated slow-motion speed, and to have two audio tracks available.

☐ *Level 2* equipment has a built-in microprocessor that allows the introduction of branching programs. In fact, this is the first real interactive level. Simple programs can be prepared on the unit's own keypad, but most programs use software downloaded from the videodisk. Participant's response is via the unit's own keypad.

☐ *Level 3* assumes the use of an external computer, usually a personal computer, as the control device. For videotape this is the only level available. For videodisk users it can be more flexible than the Level 2 package.

☐ *Level 4* is the 'anything goes' level, where extra peripherals extend the scope of the system.

Extended audio

For some users, the capacity of the videodisk is rather daunting, especially in regard to its still-frame capacity. It is clear, however, that any user wanting to make big interactive *audio*-visual programs (as opposed to an image-only library) would quickly run out of audio capacity. The available 54,000 image frames are out of balance with the 2 × 36 minutes of audio.

The lack of balance arises because the audio is recorded as a high-fidelity signal to accompany a moving-image sequence. For most interactive work this is an extravagance, so systems have been devised that allow the recording of ten seconds of audio instead of one video frame. The EECO system allows a single video image to be accompanied by up to forty seconds of audio; as shown in Figure 12.16 there is a trade-off between the number of video pictures and the still-frame audio time.

Clearly, these systems are of great interest to those who must squeeze the maximum out of one disk. For an educational or archive establishment, the idea of having one videodisk that carries 27,000 still images, each with its own ten-second commentary (amounting to 75 hours of commentary in all!) may be attractive, but this raises another point.

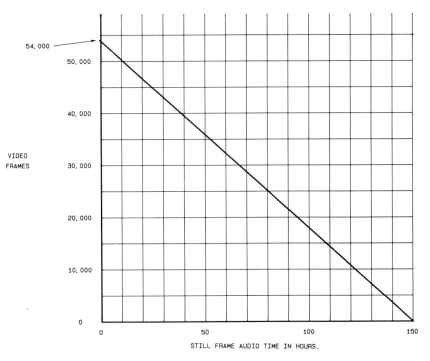

Figure 12.16 The EECO system allows a large amount of audio to be compressed onto a videodisk. The graph shows the trade-off between audio time and still-frame capacity.

The technical facilities needed to make the master tape for a videodisk side become complex and expensive if there is a lot of still-frame work, and where an effort is being made to use every available frame. This expense is not a problem to a user needing a lot of disks, but can be to the more modest user. Those requiring only a few thousand or, equally likely, a few hundred images do not need to use unnecessarily complex methods. For example, a series of slides can be recorded as a fast slide sequence, and directly transferred by a multiplexer. Their actual position on the disk can then be determined when the disk is received. A transfer rate of between one and four slides per second is feasible this way, which gives a capacity of between 2000 and 8000 images on one disk. This approach is ideal for the low-cost production of one-of-a-kind disks like the London Game example cited earlier.

Figure 12.18 Generic interactive programs work well, as shown by the promotion of microwave cooking in CWS stores. (Photo Convergent Communications Ltd)

Point-of-purchase displays

It is difficult to tell whether the biggest market for interactive audio-visual will be corporate training, or point-of-purchase displays. In fact, the two applications may well merge together. Multibranch organizations like AT&T in the USA and Lloyds Bank in the UK are already using videodisk systems for staff training and staff information at branch level. They will probably seek to make

further use of their investment by making alternative programs available to the public. Often these programs can use the same disk as the in-company program.

Point-of-purchase systems will see some of the more complex applications of the medium. AT&T is offering point-of-purchase systems that allow direct purchase via a credit card reader. However, many users will get good results from a simpler approach. Some lessons have already been learnt:

☐ Response must be fast. Programs benefit from having a presenter who is seen on the screen, but long TV-style introductions must be avoided.

☐ Programs demanding a personal input from the user hold interest. For example, in a travel program 'where have you been before?' can be asked, so that other places can be recommended. In a health-food program asking for details of age, weight etc. could be done in order to make a diet recommendation.

☐ Generic programs work well. A program on microwave cookers prepared by Convergent Communications for Britain's Co-operative Wholesale Society boosted all microwave cooker sales, not just those of the main brand featured.

☐ While true interactive programs yield the best results, they must be made with a lot of care. For many applications where budgets are limited, simple random-access programs can be a cost-effective alternative.

New developments

The power of interactive audio-visual, and in particular the contribution of the videodisk, is only

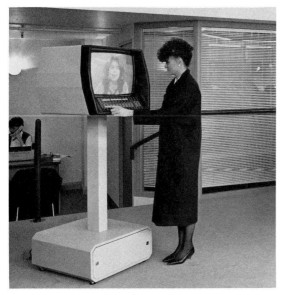

Figure 12.17 A typical point-of-purchase display. (Photo Videodem UK Ltd)

just beginning to be realized. Potential users should not be afraid to use it in a simple way; an effective simple application is better than a botched attempt at an all-singing all-dancing extravaganza. For the adventurous there are several developments on the horizon.

High definition videodisks giving 1125-line pictures with a 5:3 aspect ratio will be available almost as soon as HDTV gets seriously underway. The amazing storage capacity of the laserdisk is causing the manufacturers some difficulty in deciding how to format it. In addition to the pure videodisk, the pure memory disk, the extended audio capability referred to earlier, and the compact audio disk, new disk formats giving a mixture of facilities are available. Philips have developed a new player for Britain's Domesday Project, run by the BBC, which seeks to record details of the entire country. The data is in both computer and pictorial form.

Interaction with an audience, as opposed to an individual, is also possible. The sophistication of these systems is usually in the complexity with which the audience voting is displayed, rather than in the program itself, which must necessarily be fairly simple. Walt Disney World have an excellent example in their Communicore Theater at EPCOT, where the audience's opinion is sought on many topics, each of which is introduced by an

Figure 12.19 The BBC Advanced Interactive Video system is used in their Domesday Project. It incorporates the Philips VP415 laserdisk player which uses disks that store both video images and computer data.

AV sequence. Each member of the audience is equipped with a five-way voting panel, and survey results are shown as live computer graphics. The results are added to those of previous audiences, and are also used to determine how the particular presentation should proceed.

Chapter 13

Slide presentation systems

A review of AV media may have indicated that slides are the best medium for a particular application, but it may then seem as if the problems have only just begun. What was once the magic lantern of Victorian times seems to have become computerized. In fact, it is the very flexibility of the slide medium that has led to the enormous choice of presentation systems and formats. This chapter examines the different possibilities and defines some of the newer slide projection technical terms.

Slide projectors

All AV applications of slides use the automatic slide projector. In Europe, especially, there are many automatic slide projectors which use straight slide trays holding about fifty slides. On the whole, these projectors are not acceptable for AV use, although they are fine for the photographic amateur and for simple visual aids work in education.

Figure 13.2 The Kodak Carousel SAV series of projectors is used in Europe and 220 V countries.

The standard projector for AV work is the Kodak Carousel™. This range of projectors uses a circular slide tray containing 80 slides and operating on the gravity-slide feed principle. Two ranges are produced. For the USA and other 110 V markets Eastman Kodak produce a range using 80 or 115 V lamps. The current models for audio-visual applications are the Ektagraphic 111™ range. All Ektagraphic projectors have dissolve sockets (see later).

For the 220 V market Kodak AG of Stuttgart produce the Carousel SAV series of projectors. Currently, the main product for AV applications is the SAV 2050 projector which is fitted with a dissolve socket. There is a companion projector without a dissolve socket, the SAV 2010. Both projectors have voltage selectors for worldwide use, and both are fitted with two lamps with a simple switch change, so that important presentations need never suffer the embarrassment of lamp failure.

Figure 13.1 The Eastman Kodak Ektagraphic III projector is the standard in Canada, the USA and other 110 V countries.

Figure 13.3 The Simda projector from France is one of the competitors to the Kodak range. It features a particularly fast slide change.

Recently, Kodak AG have introduced the lower-cost 1000 series of projectors with plastic rather than metal diecast bodies. These have only one lamp and do not have a voltage selector. Only the SAV 1050 has a dissolve socket. All Kodak SAV projectors use a 24 V, 250 W lamp.

The success of the Kodak design, the expiry of the basic patent, and the widespread acceptance of the Carousel slide tray has recently resulted in other manufacturers producing Carousel look-alikes, or self-contained slide projection units which use the same slide tray. Some examples are the Singer/Telex range in the USA, the Elmo Omnigraphic from Japan, and the Simda projector from France. It should be noted that because some of these projectors have slightly different electrical arrangements, it cannot be assumed that all control equipment designed to work with the

Figure 13.4 Leitz have introduced this projector for the Carousel™ market. It features excellent optics.

Kodak projectors will work with these other projectors. It is always best, therefore, to purchase projectors and control equipment from the same source to ensure that only one vendor is responsible for the successful operation of the system.

There is a huge range of lenses available for slide projection, covering every requirement from compact back projection to long-throw projection in large auditoria. There is no excuse for having a projector perched in the middle of the room because the wrong lens is being used. The possibilities are explained in the Chapter 21.

Focusing is a common problem affecting the choice of slide projector. In an ideal world, all audio-visual users would mount their slides in glass and would set up their slide projectors properly. They would then need to focus them only once, and would have no need for either remote focus control or automatic focus.

Figure 13.5 In Europe projectors with straight slide trays are favored by amateurs and are also widely used professionally in education and training. (Photo E. Leitz Instruments Ltd)

Any professionally produced audio-visual slide show *must* be delivered in glass mounts. Any professional company providing speaker support slides for use in a company meeting or similar occasion would supply the slides properly mounted. Thus, slide projectors used for running multi-image shows and most serious applications of slide projection do not need either a remote focusing capability or an autofocus capability.

However, in the real world, both facilities are sometimes required. In a presentation room it is usual to provide remote focus control on projectors being used for speaker support slides, especially to take account of the fact that a presenter may be using slides mounted in different kinds of slide mounts. All the main

Figure 13.6 An automatic lamp changer fitted to a Kodak Carousel projector. These devices are recommended for permanent installations.

automatic slide projectors used for AV applications have the facility for remote focus control.

Autofocus is really only needed for card-mounted slides, or for plastic mounts without glass. This is because the heating of the slide in the projector gate can cause it to pop and go out of focus while being projected. Many mass-produced slides, instant slides and some bureau-produced slides are supplied in glassless mounts, usually for reasons of cost, but also because the slides are easier to mail. A user of these slides has to weigh up the importance of the occasion; if it is a major presentation, the slides must be remounted in glass. If, on the other hand, it is a weekly management meeting and the slides will be seen once only, or is an informal training session, it is reasonable to leave them glassless. In that case autofocus becomes desirable, if not essential. Only a few of the projectors widely used for AV work have an autofocus facility.

Most of the AV slide projectors using the 24 V, 250 W lamp are now fitted with two projection lamps, with a simple lever that allows instant lamp changing in the event of failure of the first lamp. For permanent installations this feature can be made automatic, either as a built-in feature of the projector (as in the Simda and Hasselblad machines) or as an external device (as in Electrosonic's ALC for the Kodak projectors). Automatic lamp changers are available from several independent vendors (e.g. GAVI) for Ektagraphic™ projectors.

What is a dissolve unit?

Most automatic slide projectors take about one second to advance from one slide to another. This means that there is necessarily a dark interval

between each slide. This may be acceptable for simple visual aids work, or for slide AV shows to individuals, but it is certainly not acceptable for prestige presentations or for AV programs being shown to groups.

The solution to the problem is to use *two* slide projectors and an electronic dissolve unit. This device controls the projector lamp brightness, and can produce a dissolve from one image to another. The basic idea is not new. No self-respecting magic lanternist of the nineteenth century would have ever given a show without dissolve picture change, although in those days the slides had to be changed by hand and the light regulated by turning gas taps.

Modern electronic dissolve units use microprocessor technology to give a wide range of facilities. In the simpler applications, slides are loaded into alternate trays and pressing the

Figure 13.7 The idea of dissolve picture change with slide projection is not new. This splendid Triunial magic lantern, owned now by Doug and Anita Lear, has hydrogen and oxygen gas regulators to achieve dissolves. It was made by J.H. Steward in 1885.

appropriate button on the dissolve unit or on a remote control gives a choice of cut, dissolve or reverse-cut slide change. The dissolve speed can usually be preset.

The simple dissolve effects are themselves quite sufficient for showing a sequence of lecture slides. However, when a sequence is to become part of an AV show with a recorded sound track, it can be useful to have a wider range of effects. Therefore, many dissolve units allow the show producer to use a lot of different dissolve speeds in one show, to deliberately superimpose slides, and to produce simple animation effects by rapidly alternating the lamps without slide advance.

Because all dissolve units need to control the projector lamp, the projector must be fitted with a dissolve socket to allow access to the lamp circuit as well as to the slide advance. Manufacturers of

AV Slide Projectors have tended to fall into line behind Kodak; on 110 V projectors using 80 V or 120 V lamps a seven-pin remote-control socket is used; and on 220 V projectors using 24 V lamps a twelvepin socket is used.

The electronic component that directly controls the lamp power is called a triac. One of the disadvantages of having this device inside the dissolve unit is that it does dissipate some heat (about 15 watts on the 24 V, 250 W lamp) and requires thick power cables from the projector on the 24 V lamp to prevent volt drop from causing a loss of light output. Therefore, some dissolve unit manufacturers put the triac in plug adaptors, which has the added advantage that the dissolve unit can be used with different types of projector just by changing the adaptor (rather than the cables).

Figure 13.8 A typical modern two-projector dissolve unit. Note the use of a simple projector stacking arrangement.

A further option is to put the triac in the projector itself. This is done by one equipment manufacturer (Electrosonic) which markets a special version of the Kodak SAV 2010. In this case a compact eight-pin control connector is used. The practice is also widely adopted by European projector manufacturers such as Leitz and Kindermann which sell mainly to the photographic, amateur and educational markets. Dissolve units made for these markets are very compact because the placing of the triacs in the projectors removes both the largest and hottest components.

A reasonable question is: why does nobody make a dissolve projector? In fact these projectors are available (e.g. Rollei). The reason why they have not found favor in the AV world, as opposed to the amateur market, is that they do not have sufficient slide capacity and do not use the Carousel™ slide tray. Also, for multi-image work it is often necessary to use more than two projectors, so it makes better sense to standardize on single-unit projectors.

Projector stands

The use of more than one projector on one screen introduces the need for special projector stands. The aim of any stand system is to get the projectors as close together as possible, while still leaving access for tray changing.

Although the rehearsal of a dissolve slide show is easiest done with the projectors side by side, the

Figure 13.9 A simple stacker stand for three-projector shows.

Figure 13.10 A precision stacker stand for permanent installations. (Photo Chief Manufacturing Inc.)

preferred arrangement for actual shows is to stack the projectors, because this keeps the lenses as close together as possible and minimizes the space taken up by the projectors.

Shows that are based on unrelated images can use simple stands. For example, if only two projectors are being used, a simple tray stand allows one projector to be placed on top of another. If three projectors are being used, a simple triple-shelf stand can be used (see Figures 13.8 and 13.9). However, if multi-image techniques, such as successive reveal or soft-edge masking (see Chapter 14) are being used, the projectors must be aligned with great accuracy.

The only practical way of achieving this is to use a stand in which the projector is held as an entity, and *not* to use its own rather crude adjustments for levelling. The stand itself has independent adjustments for each projector for roll, pitch and yaw. Stands are available that take one, two, three or four projectors.

The stacker stand itself must be raised to the correct operating height. For travelling shows fold-up stands that pack into a suitcase size are available. Multi-image shows using more than three projectors usually need a substantial shelving system. In permanent installations, the base stand should be as rigid as possible because the slightest vibration, even that caused by a slide advance on a neighboring projector, will cause noticeable picture shake. Also, in permanent installations the stacker stands should be permanently bolted to the base.

Synchronizing to sound

The principle of synchronizing slide projection to a sound tape is very simple. A separate track on the tape is used to carry control signals. Most AV shows are finally run from the standard compact cassette. This calls for the use of audio-visual tape recorders, because the crosstalk of a standard cassette recorder is not good enough to permit the use of adjacent tracks. Apart from the fact that the use of adjacent tracks would prevent the use of stereo sound, the control signal would be audible. The audio-visual tape recorder has a well separated control track which eliminates these problems, and gives inaudible control of the projection.

The educational use of single-slide projectors linked to sound is sufficiently widespread that there is an international standard for the control signal. This consists of a short burst of a 1000 Hz tone. The tone burst is detected by a simple

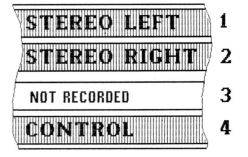

Figure 13.11 Typical track arrangement on a tape cassette used for audio-visual shows. Unlike a normal music cassette the tape only plays in one direction, and must be rewound after each show.

electronic circuit and is used to advance the projector. The same principle is used in the self-contained single-projector slide/sound units used for individual presentation and training, such as the Telex Caramate™. A 150 Hz signal is often used for tape stop or program pause.

A dissolve unit can be controlled by the same kind of signal. However, a single type of pulse signal can only give one type of command. If advantage is to be taken of multiple dissolve speeds, animation etc., it is necessary to have a more complex system of signalling.

For two-projector systems a widely used method is to have a continuous signal that is an analogue of the relative lamp intensity. Figure 13.2 shows one type of signal which uses frequency modulation. At one frequency the A lamp is full on, with the B lamp off; at another frequency the opposite occurs. Any frequency in between represents a proportional mix of the two lamps. This kind of system allows any speed of dissolve to be recorded, and, in fact, recordings can be made in real time using a slider control.

Although this method of programming is attractive to photo amateurs, and is also suitable for simple in-house production in training departments, professional shows tend to use computer methods for show programming. This results in the need to use complex digital signals for the control of the projectors. The signals used can control more than two projectors, and have the added advantage that they can include slide position data. This means that, in principle, if the tape is started in the wrong place the projectors will quickly catch up with the tape and get back in sync. The signals used are described in more detail in Chapter 16.

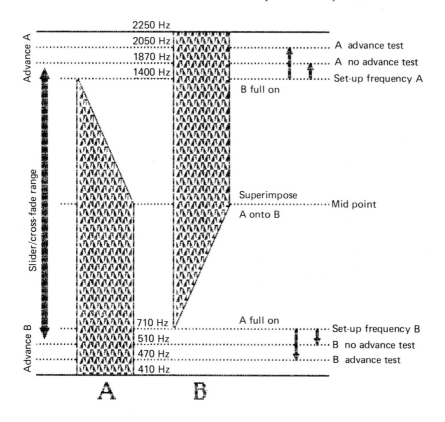

Figure 13.12 The frequency modulation system of two projector control. This is only one of several methods used.

The single-projector show

Although this chapter is mostly concerned with slide presentation systems that use two or more projectors to produce shows with a professional look, it is important to understand that there is a useful role for the slide/sound show that uses only one projector.

One of the great advantages of these systems is simplicity of use. Not surprisingly, the AV cassette recorder with a single-slide projector is widely used both in education and training. It is especially suitable when the material is not a show but, for example, a relatively long verbal description of a number of separate images.

Single-projector shows are also an effective and economic basis of individual AV communication. They give appreciably better image quality and lower production costs than any video-based system. This has led to a demand for special self-contained portable slide/sound units, either consisting of a standard projector in a special carrying case with a folding screen, or an 'AV box' construction. Some examples are shown in Figures 13.14 and 13.15.

Why three projectors?

For many users of slide programs two projectors on a single screen are quite adequate. There is no need, in fact it is an obvious disadvantage, to make a slide projection system more complex than is necessary for a particular application. However, many slide-based AV programs now use three projectors on one screen. Why?

Figure 13.13 An AV cassette tape recorder suitable for making simple shows. (Photo Kodak Ltd)

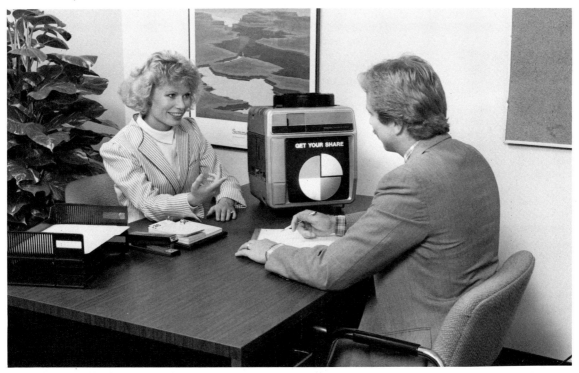

Figure 13.14 A self-contained unit for slide/sound shows, suitable for individual presentation. (Photo Gordon Audio Visual Ltd)

The main market for shows using three or more projectors is in professionally produced shows where the producer requires the extra flexibility that three projectors can give:

☐ Greater slide capacity. Shows made for sales promotion or motivation can be so fast-moving that a 240-slide capacity is needed for a twelve-minute show.

☐ Faster slide changing. Three projectors allow more rapid slide sequencing, which may not be necessary for a whole show, but may be useful for a fast-moving introductory sequence.

☐ The possibility of dissolve superimpositions, i.e. the ability to dissolve a sequence of slides on another background slide.

☐ Simpler slide production. Dissolve superimposition is one of several effects that allow the use of simple slide production techniques to produce an elaborate effect on the screen.

☐ More animation possibilities, e.g. three-way animation or two-way animation superimposed on a static background slide.

Figure 13.15 An alternative approach to the single projector show. This unit contains a standard Carousel™ projector which can be used externally or with the built-in screen. (Photo Kodak Ltd)

A show that *needs* three projectors is inherently more complex to produce and program than one that uses only two. Two-projector shows are usually programmed in real time. This means that the person programming the show listens to the sound track and simultaneously either moves a slider control or uses push-buttons to achieve the required effects. Editing is done by simply re-recording a section.

With three projectors this procedure becomes impractical. Shows using three (or more) projectors require a programming system that allows leisure time programming, so that the show programmer can enter the commands first, preview sections or the whole of the show, and then edit the commands to fine-tune the production, usually by making changes to the timing of individual effects. The process is described in more detail in Chapter 16.

The memory dissolve unit is typically designed to control three projectors. The activities of each of the three projectors can be programmed independently, with a selection from a wide range of dissolve speeds and animation effects. In addition, the timing between individual cues or groups of cues can be pre-programmed, so visual effects can be precisely rehearsed, and, once completed, the show producer can be sure that each effect will be played back exactly as intended.

Memory dissolve units can store about 1000 individual cues, which is more than sufficient for the average three-projector show, even if all the slide trays are completely full. Completed shows

can either be run from the dissolve unit's memory or, more usually, can be transferred to magnetic tape in synchronization with a sound track. The dissolve unit then acts as a playback device, decoding the complex signals from the control track of the tape. For users only needing to play back ready-made shows, three-projector dissolve units are available without the programming capability.

A dissolve unit does not necessarily control either two or three projectors. Some vendors offer unit dissolvers where there is one dissolve unit per projector; others offer dissolve units able to control four projectors. However, after the simple two-projector dissolve unit, the three-projector unit has become the most common, both because the three-projector stack is the most common and practical multi-image unit, and because it represents an economic use of microprocessor power.

The presentation unit

Most end users of slide-based audio-visual are more concerned about the successful showing of completed shows than about the mechanics of their production. Even at two projectors, let alone three, the system is already beginning to look a

Figure 13.16 A three-projector memory dissolve unit. This device can program complex shows, and has a large built-in memory to allow sequences to be built up at leisure.

Figure 13.17 A two-projector audio-visual presentation unit. This unit can both make and play back shows, and is suitable for quite large audiences. Units of this kind are widely used in training departments.

Figure 13.18 A three-projector audio-visual presentation unit. This heavy-duty unit is intended for the playback of professionally made shows with high-quality stereo sound. It is possible to add more three-projector dissolve units for shows using more than three projectors.

little complicated, especially because it includes separate projectors, dissolve units, AV tape recorder, amplifier, etc. Obviously for most users it would be best to combine all the electronics into one box.

This is done with the audio-visual presentation unit. Two types of unit are common. The first is a unit intended for training departments and similar users who wish to make their own, relatively simple, shows. A typical unit is a combination of an AV cassette recorder, power amplifier and two-projector dissolve unit.

The second is a playback unit only, suitable for either two or three projectors, and intended for the playback of professionally produced shows. These units usually include high-quality stereo sound reproduction with an amplifier power of at least 2 × 20 watts. They should have some system of noise reduction if they are to be used for large audiences. Most allow for the addition of extra dissolve units for running larger multi-image shows.

Presentation units are usually intended to be used for occasional and travelling shows. Not all are suitable for permanent installation, so users requiring such units for fixed or continuous-running shows should check on the suitability of a particular unit for unattended use. Additional features of these units are:

☐ The unit is fitted with a heavy duty automatic tape deck.

☐ The unit has the facility for controlling the projector power for automatic switch-off in unattended push-button started shows.

☐ The unit has built-in routines for continuous running or one-shot shows, with full auto-present routines to ensure that projectors are checked as set at 'home' before the start of each show.

Why multi-image?

It is always a matter of debate as to where single-screen slide/sound stops and multivision or multi-image starts. It is usually defined as more than three projectors. (The word multivision is essentially European and is interchangeable with its American equivalent, multi-image.)

Thus, it is possible that multi-image will have yet more projectors on one screen, or, more likely be multi-screen. The reason for requiring more projectors is simply an extension of the arguments for three projectors: higher speeds, and more elaborate effects. The reasons for more than one screen can be summarized as:

☐ To match the show to the environment. A single standard-format screen can look singularly unimpressive in a large auditorium. The multi-image technique allows the creation of big screens to match large audiences.

☐ The low ceilings of many conference sites demand a long, low format.

☐ To allow the creation of a total show environment by using the flexibility afforded by multiple projection.

☐ To produce special show formats, especially in exhibitions and in permanent installations in museums, visitors' centers and tourist attractions.

Because the multi-image medium is so flexible, there are not really any set rules about screen formats. However, it is natural that many users will have similar requirements, and there are several popular formats (see Figures 14.4, 14.8 and 14.9). The majority of multivision shows made for business presentations are in the nine to fifteen projector range, with only an occasional requirement for bigger systems for exceptionally large audiences or sophisticated shows.

Some public entertainment and exhibition shows, on the other hand, use very large numbers of projectors. Shows using 48, 60, 96 and 120 projectors are not common, but neither are they unknown. Further information is given in Chapters 14, 15 and 16.

Figure 13.19 A 42-projector permanent installation. A clean projection room with easy projector access is essential for these systems. (Photo Electrosonic Systems BV)

Figure 13.20 A specially designed set incorporating multi-image slide projection. (Photo of their 25th anniversary presentation courtesy Swedish Steel Corporation)

Staging slide shows

One of the problems of slide-based AV shows is the number of separate items involved. Despite the underlying simplicity of the slide medium, there can be a tendency to underestimate the effort needed to ensure a perfect show.

There are few problems associated with the extremes of the business. The single-projector show, even the simple dissolve set-up, presents little difficulty. At the other extreme, the large permanently installed system, provided it has been correctly engineered and receives the recommended routine maintenance, will give reliable continuous service.

The problem is in the middle, where, for example, an industrial user has a six-projector show which he wishes to move round several different venues. It can be tempting to understaff the exercise, and involve the presenters in the business of staging it. A good rule is never to allow anyone who is part of the presentation to touch equipment!

This suggestion applies to any kind of presentation using AV (not just slide-based), but is especially important with slide and video projection because of the care that is needed to align the projectors, and the know-how to avoid environmental problems in strange hotels. Therefore it is best to have a technician, or at least someone who regularly uses the same equipment set, responsible for staging the show.

Chapter 14

Slide formats, mounts and masks

Although in theory there are a large number of slide formats, in practice the situation is now very simple. The great majority of slides are made on 35 mm film, and it is the dimensions of this film that determine what can be carried on a slide.

Slides are mounted in 50.8 mm (2 in) square mounts, but of course produce a rectangular picture. This led to the development of the superslide which has an opening of about 38 mm (1.5 in) square, which is the largest possible within the mount. Superslides are made either by cutting down transparencies made on 120 film (i.e. from 54 mm square) or, more usually now, by rephotography onto a special 46 mm film on a rostrum camera. This film is perforated only on one edge and can only be used in special cameras.

Sometimes it is necessary to project original transparencies of a larger format. For this purpose there are alternative mounts 70 mm (2.75 in) square, which allow the direct mounting of 54 mm square transparencies. The application for these slides is mainly in the field of professional photography, top amateur photographers and advertising agencies. Even larger sizes are used for special purposes, such as movie advertising and for some projection in theaters.

All slides should be properly mounted for projection. Although there is an argument for using glassless slide mounts in some high-power projection applications, all normal applications *must* use glass-mounted slides. Glass mounts protect the slide, allow for cleaning when necessary, and ensure that the slide remains properly in focus when projected. There is no place for autofocus in professional slide projection.

Most AV slide mounts use anti-Newton ring glass. 'Newton's rings' are colored fringes produced when light passes through two surfaces very close together, which can happen when film and glass are placed next to each other. If there is a gap between the two which is close to the wavelength of light, the colored rings are seen. Anti-Newton Ring (ANR) glass has a roughened surface to ensure that there cannot be a significant area of film that is the critical distance from the glass.

For all normal AV use, mounts with ANR glass are not only acceptable; they are essential. However, purists would point out that areas of film that are projected as saturated white allow the ANR coating to be seen as a mottled effect when viewing the screen from close up. Thus, some super-critical photographic applications call for the use of ANR glass on the film backing side and plain glass on the emulsion side. The argument here is that the emulsion itself can meet the ANR requirement, and because focusing of the slide is on the emulsion side, the ANR glass is not seen.

The degree of roughening of ANR glass does vary between glass manufacturers, so the recommendation is to carry out a viewing test if it is thought there could be a problem.

Slide mounts are designed to slightly mask the original image, by about 1 mm, to ensure a clean edge and to allow some tolerance in the mounting.

Figure 14.1 A range of dustproof slide mounts: 35 mm, Superslide, 54 mm square and Idealformat. (Photo Perrotcolor)

An audio-visual show will never use original slides as show copies, both for security reasons and because colors deteriorate with projection. Slide mounts with an extra large aperture are available, therefore, to ensure that when an original slide is duplicated there is no further loss of image area. These are known as duplicating mounts and typically have an aperture of 35.8 mm × 24.4 mm.

Table 14.1 Standard slide formats

35 mm

Mount size	50.8 mm × 50.8 mm
Image size	36 mm × 24 mm
Available aperture	35 mm × 23 mm

Superslide

Mount size	50.8 mm × 50.8 mm
Image size	40 mm × 40 mm
Available aperture	38 mm × 38 mm

54 mm or 2¼ square

Mount size	70 mm × 70 mm
Image size	54 mm × 54 mm
Available aperture	53 mm × 53 mm

Note that available apertures are approximate. They represent the masking provided by the mount and vary from make to make. See also Chapter 21

Registration mounts

To simply project a sequence of photographic images on a single screen, there need be nothing special about the slide mount provided that it securely holds the film, and the complete image is projected.

However, many AV applications have much more critical requirements. For example, successive-reveal slide sequences or build-up graphics look amateur if the base image seems to jump around the screen. These sequences, and any multi-screen work where one large image is made up from a number of separate slides, require great precision in slide mounting.

Just getting the mounting right is not sufficient. The whole process of slide production must keep the image in register from the start, i.e. the image position on the film always bears a fixed relationship to the position of the film sprocket holes (see Chapter 2). Finally, the slide must be mounted in a register mount which has pegs that accurately locate the film.

There are several reputable makes of registration slide mounts. The lower-cost mounts are

Figure 14.2 A register slide mount. The film sprocket holes are accurately located by the molded pegs. (Photo Perrotcolor)

two-part and require assembly. Business users prefer the register mounts with hinges; although more expensive, they are very quick to use. For ultra-critical applications it is possible to specify that a batch of mounts is all made from the same mold.

Sometimes it is necessary to register a piece of film that has not been shot on a register camera, or for some reason is out of alignment with other images in the sequence. A viewer/punch is available that allows the user to align a slide and then punch round location holes in the side of the film. A special variable-registration mount, with location pins to match these holes (instead of the usual sprocket pegs), is then used. In all other respects the mount matches the normal register mount.

This variable-registration approach is also useful for those who have little general need for registration and are able to do most of their work on non-register cameras. The variable-registration mount can then be used for the few slides in a sequence that are critical.

Slide masks

Slides used in multi-image shows can be a long way from a standard 35 mm photo. The rostrum camera, with its ability to manipulate original material and to give accurate multiple exposure and many other tricks, has transformed the humble magic lantern slide. However, not everything is done with the camera. Multi-image shows

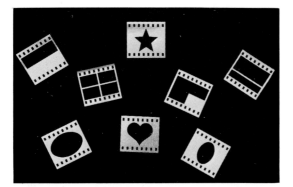

Figure 14.3 Aluminum masks used for special effects. These have the advantage of being completely opaque, whereas photographic masks on high-contrast film need careful inspection to avoid pinpricks of light. (Photo Perrotcolor)

make frequent use of masks sandwiched in the slide mount with the film. These masks may either be thin metal or high-contrast lith film.

Thin metal masks, made of aluminum or stainless steel, have the advantage that they are completely opaque. Film masks have the advantage that *any* shape can be inexpensively made as a mask. However, they still transmit some light and before use must be examined for scratches or small imperfections that might give pinpoints of light on the screen. These can be covered using an opaque compound.

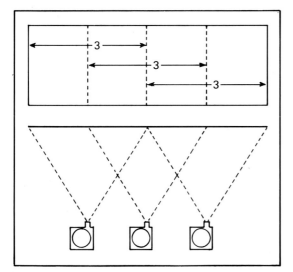

Figure 14.4 Probably the most popular multi-image format uses nine projectors on a single 3:1 ratio screen. Two images side by side with a center-overlap image use soft-edge masking to achieve single panorama pictures.

One particular kind of mask is widely used in multi-image to produce large images from a number of slide projectors without any dividing lines. Normally attempts to produce this kind of image fail because it is impossible to line up the projectors with enough accuracy so that the viewer is unaware of the picture join. The soft-edge masking technique, however, eliminates the problem.

The technique is best described by an example. Probably the most popular multi-image format is an overall screen ratio of 3:1, achieved by projecting two standard slide images side by side. How, then, can a single image without any apparent join be achieved? It is done by projecting a third image on the center of the screen, with the center projector projecting half of each of the pictures projected by the left and right projectors.

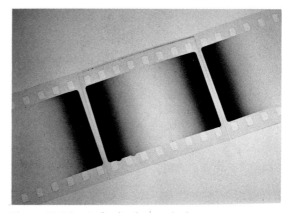

Figure 14.5 A set of soft-edge masks for panorama projection. (Photo RCW Colour Slides Ltd)

In order to achieve uniform screen illumination each slide is fitted with a soft-edge mask. For slides from the left projector, the mask is completely transparent over the left half of the picture, and gradually gets darker towards the right-hand edge, where it is opaque. The arrangement is reversed on the right-hand projector. The center projector's mask is completely transparent in the center, and opaque at both edges. The mask sets are matched so that the overall picture is uniformly illuminated.

This technique only works if the projectors are precisely lined up, and if the slides have been produced in proper register with each other. Even wider screens can be used by repeating the process. The technique has even been done in the vertical direction to produce very big images. It is not actually necessary to use a 50 per cent overlap;

Figure 14.6 A soft-edge panorama at the Royalty and Empire Exhibition, Windsor, England.

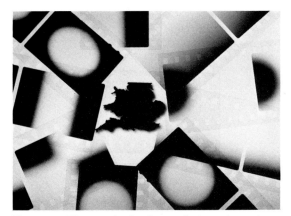

Figure 14.7 The use of soft-edge masks is not limited to panoramas. They can be used for montage and vignetting effects. (Photo RCW Colour Slides Ltd)

a 25 per cent overlap can be sufficient. However, fifty per cent is often preferred because it gives more flexibility in the use of the projectors.

Show formats

Single-screen shows using slides normally use the standard 35 mm slide in the 'landscape' format. This gives a satisfying 3:2 ratio. No professional AV show mixes 'landscape' and 'portrait' format slides.

Occasionally superslides are used on a single square screen, but this format is not very satisfactory, because if the screen height is fixed by practical considerations, usually a low ceiling, the available picture can look smaller. Superslides can, however, make a most satisfactory component of multivision, because some producers prefer the square image and all the available light from the projector can be used.

Multi-image formats are a matter for the imagination – anything goes! However, because many business programs must be shown in offices, hotel banqueting suites and other places with low ceilings, there are a number of preferred formats. On the other hand, shows for exhibitions and display do not need to suffer from such restrictions, and the designer can be adventurous.

Although business shows, and shows of a documentary nature used in visitors' centers, make a lot of use of the single big screen with the soft-edge masking technique, exhibition shows often make use of the 'matrix screen' approach. Here the individual images are deliberately divided, and the arrangement of the divisions is part of the display design. Usually the screens in these displays are relatively small in order to ensure adequate screen brightness.

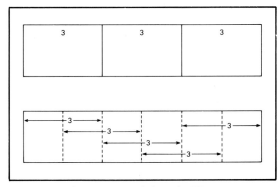

Figure 14.8 Three images side by side. The screens can either be deliberately mullioned, or more projectors can be used with soft-edge masking.

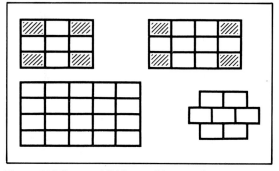

Figure 14.9 Some exhibition multi-screen formats. Here each image would be quite small. The shaded screens are optional.

Chapter 15

Special slide projectors

The needs of the great majority of applications of slide projection are met by the standard automatic slide projectors using 35 mm or superslides. Some applications, however, call for special equipment because of the need for more light or special control. Some of these special requirements are reviewed in this chapter.

More light

The most common request is for more light, either because the user wants a bigger picture, or to fight ambient light conditions. There are ways of achieving brighter images, but an understanding of the basic problems involved will help to determine whether a high-output projector should be used or whether there is some alternative solution to the problem. Unfortunately, there have been cases of users being disappointed by the performance of bright projectors because they have not understood the fundamental problem.

The human eye is good at comparing light levels on a side-by-side basis, but not on an absolute basis. Its response is not linear, so if the light output is increased by, say, 30 per cent, it is not very impressed. A *doubling* of light output is noticeable, but to create a real impression the light output must be increased by a factor of four.

A standard slide projector can give a good 4 m (13 ft) wide picture in a darkened auditorium. However, a large auditorium might need an 8 m (26 ft) screen to ensure the correct picture size for the back row of the audience. A doubling of the picture width means a fourfold increase in picture area. To get the same screen illumination a projector with four times the standard light output is needed. This application is one where a high-power projector is justified, because many users of convention auditoria want to show single slides to large audiences.

However, if the multi-image technique is being used in a large auditorium, a number of projectors can be used to create the larger images required; perhaps surprisingly, this approach can actually be less expensive, as well as giving much greater flexibility. Thus, a large auditorium should be *permanently* equipped with a high-power projector pair for lecture purposes. When individual sales conferences make use of the auditorium to run automatic AV shows, they will probably be better served using conventional projectors with multi-image technique.

Tables 15.1 and 15.2 give an idea of the different light outputs and image sizes of various standard

Table 15.1 Image sizes from a standard slide projector

This table is for guidance only; the actual image achievable will depend on site conditions. The dimension is image width on 35 mm or superslides.

Front projection in darkness	5 m	(16.4 ft)
Rear projection in darkness	3.5 m	(11.5 ft)
Rear projection in low ambient light	2 m	(6.6 ft)
Rear projection in medium ambient light	1 m	(3.3 ft)

Table 15.2 Approximate light outputs from slide projectors

The figures are for 'Open Gate' with a superslide aperture, using a 60 mm f/2.8 lens. These are approximate figures for total lumens reaching the screen. It is important that projectors give an even distribution of light, especially in multi-image applications.

Kodak SAV 2050	1000 lumens
Kodak SAV with 400 W lamp (average)	1800 lumens
Kodak SAV with 400 W and f/1.2 lens	2500 lumens
Ektagraphic 111 with regular lamp	1000 lumens
Ektagraphic 111 with short-life lamp	1800 lumens
ORC Xenographic 500	4000 lumens
ORC Xenographic 750	6000 lumens

and special projectors. Some shows might mix screens of different sizes and could justify a mixture of projectors, in order to achieve uniform illumination on the total screen area.

A point frequently overlooked is that there are other ways of achieving subjectively brighter images. A photographer might argue that a particular slide looks best when exposed one stop down, but this halves the light! It is far cheaper to double the light on the screen by judicious choice of slides than to use expensive projectors. Graphic slides with bright colors (no deep blues or dark purples) and selective rephotography with just one stop more can help a lot. On the equipment front, dirty optics, unaligned lamps, low electricity supplies and the use of non-multicoated projection lenses can all cause less light on the screen than is really available.

How is more light achieved? One way is simply to use a bigger lamp. For example, there are several conversions of the SAV projector that use a 400 W lamp. On the Ektagraphic projector it is possible to use a short-life lamp that gives a useful increase in brightness. This approach is particular-

Figure 15.1 The Optical Radiation Corporation Xenographic xenon arc slide projector uses a compact xenon lamp module and is fitted with a mechanical fade device.

ly suitable to help equalize screen illumination when different size screens are being used, or where the screen size is marginally too big for the regular lamp. It is acceptable for conference work and other one-off shows, where the procedure can be to use standard or under-run lamps for rehearsal , and a brand new set of lamps at full power for the actual show.

This type of approach is economic for single conference applications, but is clearly no good for permanent installations, or where more than double the standard light output is required. Here, the only realistic choice is the use of projectors with xenon arc lamps. Although expensive, they do give a significant increase in light output, and lamp life is respectable (2000 hours on some models). A xenon lamp cannot be directly dimmed, so it might be thought that these projectors cannot be connected to a dissolve unit. However, projectors are available with motorized fade modules which allow the direct connection of standard control equipment.

The xenon arc has a different 'color temperature' from the quartz halogen lamps used in the standard projectors. It makes the slides a little bluer and subjectively they appear even brighter. It may be necessary to take care when mixing standard and xenon projectors.

The limit to the light output from the projector is the luminous flux through the slide itself. Even with a perfect projector that puts only light radiation through the slide and rejects all infra-red heat, there is a limit. This is because a slide works by selectively absorbing different wavelengths of light, and this absorption is turned to heat within the slide. The most that a superslide can take is about 7500 lumens. In the theater where very large and very bright single images may be needed for scene projection, big slides up to 25 cm (10 in) square are sometimes used in very special and very expensive projectors giving up to 50,000 lumens!

Slide life decreases more or less in direct proportion to the light passed through the slide. Users of high-power projectors in permanent installations must remember to allow for extra slide copies accordingly.

Some users have tried to use xenon arc slide projectors for public display in shopping malls. This can work well, but only if close attention is paid to ambient light conditions. *No* projector can defeat sunlight, which can be thousands of times brighter than the brightest of projectors. Such systems *can* be a great success if properly designed, but can equally be a disappointment if realities are not faced at the outset.

Special-format projectors

The standard 35 mm slide and its near relation, the superslide, meet the needs of most users of slide projection. However, photographers, art directors, advertising agencies and others sometimes need to project slides of a larger format, either because they are seeking the ultimate in quality, or because they wish to project an original transparency.

Figure 15.2 The Hasselblad PCP80 Projector for 54 mm × 54 mm slides.

This has traditionally been done using either manually operated projectors, or ones of an amateur construction. Recently Hasselblad has introduced an automatic slide projector for 54 mm × 54 mm slides. It is built on similar principles to the Kodak Carousel, and has many features that make it suitable for professional users.

This type of projector can be the basis of very spectacular audio-visual shows. However, there is an important point to bear in mind. The ability of the human eye to resolve detail is amazing, but not so amazing that it can pick out fine detail at great distances. The benefits of this kind of projector will be greatest when the viewer is relatively close to the screen, and is able to appreciate the improvement in image quality.

Three-dimensional projection

It is possible to project three-dimensional slide images using the principle of polarized images. The original photograph is taken by a special double camera, or two cameras with their lenses separated by a distance similar to that between two eyes. The two slides are then projected by two projectors, each with a polarizing filter. The polarizing of light can be likened to making all the light waves move in one plane, like a rope shaken to produce 'waves' but only permitted to go through a vertical set of railings. Thus, light can be horizontally or vertically polarized. If polarized light is viewed through a polarizing filter, the amount of light seen will depend on whether the light is polarized in the same direction as the filter.

Theoretically nothing is seen if horizontally polarized light is viewed through a vertically polarized filter, and vice versa. In three-dimensional slide projection two slide projectors are used to project each image. One of them has a horizontal polarizing filter, the other a vertical polarizing filter. The audience view the show through a matching pair of polarizing spectacles, thus ensuring that each eye only sees the image meant for it.

Three-dimensional slide projection is done with standard projectors, with the sole addition of the necessary filters. If dissolve equipment is being used it is necessary either to double up on the control equipment, with units working in parallel pairs, or to use dissolve units that have been specially modified to have dual outputs. It is advisable to use a screen designed for three-dimensional work, because some screens can depolarize the incident light.

The making of three-dimensional slide shows is, in principle, no more difficult than making ordinary two-dimensional shows. The discipline of planning, scripting, storyboarding etc. is the same as usual. Simple exterior shots are not a problem. However, when it comes to trick effects, especially 'matting' a foreground object into a background, a great deal of experimentation may be necessary to achieve convincing results.

Random-access slide projection

Most users of slide projection show the slides in a predetermined sequence, so the slide projectors simply advance from one slide to the next. For some applications it must be possible to select slides at random from a library, so that any slide can be retrieved in, say, three seconds.

Projectors providing this type of facility are called random-access projectors. They require some kind of motor drive mechanism to drive the slide magazine rapidly to the required position. European Kodak offer a special version of their SAV series of projectors, the S-RA, which has an external drive unit. Some other vendors offer clip-on attachments for the standard projector. In the USA several vendors offer suitably modified

Figure 15.3 An exhibition display using random-access slide projection. This display also uses random access sound. (Photo Electrosonic GmbH)

Ektagraphic™ projectors. The Simda projector is unique in that it is already designed for random-access work, provided suitable remote controllers are used.

The possible uses of random-access slide projection are many:

☐ The display of maps, building drawings, system drawings, plant diagrams etc. in control rooms, emergency operations centers, control consoles and other industrial and public service environments.

☐ In briefing, presentation and seminar rooms, where slides, charts, maps etc. may need to be shown to groups on demand.

☐ In product display applications, both in showrooms and at the point of sale.

☐ In simulators and other high-level training applications.

☐ In museum, exhibition and visitors' center applications.

The method of selection varies according to the complexity of the application. Usually single projectors are operated by a simple calculator-style keyboard, but frequently projectors must be operated by computer. In this case the random-access projector can be supplied with an RS232 interface to allow direct connection to the computer. On other occasions the activities of several such projectors must be coordinated, and this can be done either by computer control or by special control panels.

Simple random-access projectors can be purchased from an AV dealer. However, special applications usually call for an element of project engineering, and may need direct contact with the manufacturer. Some of the many technical possibilities that now exist include the automatic display of sequences. For example, a museum display might give a visitor the choice of several slide sequences on different subjects. Pressing the appropriate button immediately selects the first slide in the sequence, and the remainder are shown at a preset interval. At the end of the sequence the screen reverts to a 'home' slide. The arrangement can be extended to include random-

Figure 15.4 A presentation room control panel incorporating random-access selection of slides.

access sound, so that a single display is able to give a choice of several mini-programs, all with nearly instantaneous start. These systems are not limited to single projectors; both simple dissolve and full multi-image facilities can be added.

Random-access slide projection can be combined with computer data display. For example, a monochrome high-definition data projector can

Figure 15.6 The Simda 2200/400 projector uses a 400 W tungsten halogen lamp. There is provision for deliberately over-running the lamp to give still more light at the expense of lamp life.

Figure 15.5 Here a computerized information display is linked to a random-access slide projector. Slide projection has been chosen to give the large high-quality images (in this case of flowers) that could not be achieved by video. (Photo from Kodak Ltd)

superimpose on a high-quality color photographic image projected by the random-access projector.

The choice of which kind of random-access imaging system to use will always depend on the final application, and on the way in which the software is going to be produced. For individual instruction on a semi-institutional basis the video-disk is technically the most attractive method, whereas for storing and retrieving thousands of fingerprint records, random-access microfiche may be the answer. Random access slide projection succeeds in those applications where slides are the preferred software, for example where:

☐ Big images are needed.

☐ Photographic quality is needed.

☐ Individual images must be easily and inexpensively updated.

☐ A single-venue application is involved.

In all these cases a slide system will be the most cost-effective and simplest system to install.

Chapter 16

Multi-image programming

This chapter reviews the subject of multi-image programming. Some may think that this is too specialized to be of interest to the person commissioning audio-visual programs, but a knowledge of the basic process involved is of help in understanding how any AV sequence is put together. Although this discussion centers on multi-image as it is currently understood, based mainly on slide projection, similar principles will apply to any video multi-image systems that may emerge in the next few years.

Multi-image is such a flexible medium that at first it is a little daunting to try and analyze what is going on. But however complex the format, whether a show is using six projectors or sixty, a multi-image system is completely modular. Thus, in principle, it is only necessary to understand how a small system works to be able to design programs for any size system.

System basics

Consider the nine projector system shown in Figure 16.1. From a theoretical point of view, it does not matter how the projectors are arranged. They could each be directed at a separate screen (nine screens, one projector per screen) or even all be directed at one screen (one screen, nine projectors). In fact, they are likely to be organized as three groups of three, because for the reasons explained in Chapter 15 a group of three projectors is able to give a wide range of effects.

The requirements of a programming system are very simple. Only a few parts of the slide projector can be controlled:

□ the lamp;

□ forward slide step;

□ reverse slide step;

□ the projection of a slide specified by number;

□ possibly, the detection of the zero position of the slide tray;

□ possibly, separate electrical control of the lamp shutter;

□ possibly, control of electrical power to the projector.

The only one of these items requiring any elaboration is the lamp. It should be possible to fade the lamp on and off at a choice of speeds, ranging from a quick cut to a long, slow dissolve, and to stop the lamp at some specified intermediate brightness. Lamp flashing and animation effects are simply achieved by a rapid repetition of fast lamp on and off commands.

The job of the programming system is to organize the actions of a group of projectors with respect to time and to each other. All multi-image shows require careful planning. If a proper storyboard is prepared, the task of programming the show is relatively simple. In the old days programming was time-consuming, not so much from the point of view of entering the initial program, but more because of the difficulty of editing and making subtle timing adjustments. The arrival of the microprocessor made multi-image a more accessible medium, with the side benefit of allowing a wider range of effects.

For small shows using only three projectors it is practical to have the entire programming system in one unit, the three-projector programmable dissolve unit described in Chapter 13. For more than three projectors it is better to have a split system where there is an interface local to the projector which is, in turn, controlled by a central programmer. This type of arrangement simplifies the wiring, and keeps the high-current lamp control local to the projector. Most systems endow their interfaces with considerable local intelligence, greatly reducing the load on the central programmer.

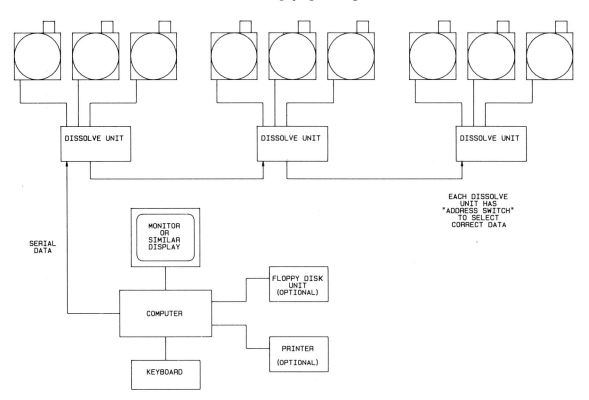

Figure 16.1 Block diagram of a typical programming
set-up for a nine-projector show.

Thus, from commands issued by the central
programmer the interface could be expected to:

☐ Fade the lamp at any specified speed. Some systems
offer a limited repertoire of speeds, but typically can
offer any speed between 0 and 99.9 seconds specified to
0.1 second.

☐ Stop the lamp fade. Some systems offer this as a 'stop
fade' command; others allow the precise specification of
the required lamp intensity, e.g. any level between 1 per
cent and 100 per cent specified to the nearest 1 per cent.

☐ Advance the projector to the next slide. To simplify
programming this is usually done automatically when
the lamp fades to extinction, with a separate 'no step'
command if, for some reason, this is not required (e.g.
for an animated sequence).

☐ Select a specified slide by number. This may be
required at the beginning of a show if one slide tray
holds the slides for more than one show, or to allow for
different versions of the same show, e.g. shortened
versions or different language graphics.

☐ Select an animation routine. An animation routine is
set up by causing the lamps in different projectors to

flash on and off sequentially. This can be done by
inserting a loop in the main program, but can also be
done by delegating the task to the interface unit. Some
systems allow the choice from a library of standard
effects, others use a system of downloading an
animation sequence from the main programmer.

☐ Operate the lamp shutter. The filament lamp used in
slide projectors (especially the low-voltage lamp used in
European projectors) has an appreciable thermal inertia.
This means that when it is switched off the brightness
decays over an appreciable time, limiting the speed of
the visual effect. Some systems allow access to the
projector shutter to give an instantaneous 'snap' picture
change.

☐ 'Home' the projector. In unattended automatic shows
it is important to be able to return all the slide trays to
zero or 'home' at the end of the show. The interface unit
should check that the trays really are home by means of a
microswitch that detects a notch in the slide tray
identifying the zero position.

☐ Switch power to the projector. Not all systems allow
this, but it is a convenient feature, especially for
simplifying the installation wiring in permanently
installed systems.

Figure 16.2 A typical projector controller for multi-image. This one controls three projectors independently. It can replay two different signals from tape (Alphasync™ and Positrak™) or can be directly addressed by a computer via an RS232 link.

Some interface units boast additional features that may make them suitable for special applications. These include power fail memory, a system that ensures that in the event of power failure the unit remembers exactly where the projector is, so that when power is restored there is no wasted resetting time; and lamp failure detection, a feature not so necessary now that projectors in permanent installations can be fitted with automatic lamp changers.

The arrangement of interface units varies between systems. At one extreme there is one interface unit for each projector; at the other, one interface unit for every four projectors. The most common arrangement is for one interface unit for every three projectors. This arrangement is due to a combination of economics and practicality. The three-projector stack remains the most common component of multi-image configurations. Note, however, that the fact that one interface controls three projectors does *not* mean that they are linked in control terms. Each projector is still individually addressable.

Usually a data cable links all the interfaces in a system, and each interface is fitted with an address switch that identifies which unit it is in the chain.

The central programmer is another microprocessor-based device. It can either be a 'dedicated' device, that is designed only for the creation of multi-image shows, or it can be a general-purpose microcomputer. In the first case the unit is fitted with a keyboard that is specific to multi-image programming, a simple extension of the controls provided on a programmable dissolve unit. However, the task of programming and editing a multi-image show is one that is almost

ideally done by a small computer, and today most systems are based on the use of personal computers. Typically, these will either be of the IBM PC™ compatible variety or of the Apple IIE™ variety. While in principle any small computer could be used, a fully developed computer program for multi-image show creation is a very complex piece of software and is not necessarily easily transported from one kind of computer to another.

Multi-image does place some special demands on the computer with regard to communications. For this reason, the computer is either specially configured for multi-image work, or requires the addition of one or two accessory cards that plug into spare slots in the computer chassis.

Using the microcomputer as a multi-image programmer is very similar to using it as a word processor. Most people *can* use a typewriter or word processor, although they may be rather slow at it. Similarly it takes some time to become a *skilled* programmer and able to work very fast, equivalent to touch typing, although most users do not necessarily have to work at high speed. The most important thing is to be able to think logically how a picture sequence is built up, and to

Figure 16.3 Multi-image programming in progress.

understand both the power and the limitations of the programming system. It is interesting to note that many of the major producers of multi-image presentations do not have expert programmers on their staff, but rely on the services of highly-skilled freelance programmers when they want a large complex show programmed quickly. This just reinforces the point that the acquisition of a basic understanding of programming is accessible to anyone, but the need to reach the equivalent of '100 words a minute' is limited.

Like a word-processing program, the multi-image production program has many editing features designed to simplify the task of initial entry and subsequent editing:

☐ Shows are finally stored on floppy disk. They can be stored as whole shows or part shows. Whole shows can be retrieved for editing or showing. Part shows can be retrieved to be added on or inserted into a show being worked on.

☐ Cues can be modified, either individually or in groups. They can be deleted either selectively or in specified blocks. Extra cues can be inserted between existing cues.

☐ Plain text remarks can be included in a show listing to make it easier to find a particular place. A tab facility is used to simplify moving to a point in the show whose cue number or timing may have changed as a result of editing.

☐ Show listings can be printed as hard copy using a standard printer.

Show principles

The easiest way to understand the process of multi-image programming is to talk through what happens in a typical show. Reverting to the nine-projector example, it is convenient to make the assumption that the projectors are arranged in three groups of three, each aimed at a different image area or screen. The image areas may well overlap, but this in no way affects how the program is made.

The program can be considered as a sequence of events, so it is easiest to start by giving these events a number. Thus, a show consists of a number of cues, sequentially numbered, usually from zero. When evaluating different systems, it can be confusing to evaluate the capacity of the system; some systems claim many thousands of cues, others are more modest. In fact, closer examination usually reveals a marked similarity in *real* capacity. In order to make the show-making program easy to understand, and to make economic use of the computer's memory, most

systems have some limitation in the amount of information stored in one cue. So with some systems it may be necessary to use two or three cues to achieve what can be achieved with one cue in another system. Now that computer memory is so cheap there should be no serious constraint on the number of cues available, and most systems allow several thousand cues to a show.

All programming systems use a kind of short-hand to limit the number of keystrokes needed to achieve a particular effect. The listing of the first few cues of a typical nine-projector show is shown in Table 16.1. While the actual command designations are those of one particular system, those of other systems are, in general, just as easily understood.

Table 16.1 Typical cue listing for the start of a nine-projector show

000	ALL	STEP	
001	F3	1A 2A 3A	
002	D2	1	T1.5
003	D2	2	T1.5
004	D2	3	
005	D1	1 2 3	
006	A510	2C	
007	F0	2C	
008	F0	2C	
009	D2	1 2 3	
010	L04		
011	D0!	1	T0.3
012	D0!	2	T0.3
013	D0!	3	T0.3
014	LE		
015	**	SUNRISE SEQUENCE	
016	D5	1 2 3	

Cue 000 can be considered a housekeeping cue. It is good practice *not* to have a slide in the zero position, because it is easily lost. Thus, the first slide to be seen will be in position 1. In this example the projector groups are designated 1, 2 and 3, and the projectors within a group A, B and C. Thus cue 001 is intended to bring up an image simultaneously on all three image areas.

The code F3 1A 2A 3A means: 'On projectors 1A, 2A and 3A, fade the lamp up at the three-second rate'. Thus, F5 would designate a five-second fade, F6.6 a 6.6-second fade and F76 a 76-second fade.

A dissolve change is when one image fades into the next. It would be possible to make the next cue the command F2 1A 1B, which would mean: 'Fade projector lamps 1A and 1B at the two-second rate'. F can be used whether the lamp is going up or down; if a lamp is down it will go up, and vice versa. However, although it will be necessary to use the F command or its equivalent if images are required out of sequence, or when doing superimpositions, the F command is a little clumsy if only a conventional sequential dissolve is required. Cue 002 is therefore shown as D2 1, which means: 'Do a two-second dissolve on screen 1'.

The D command moves round a dissolve loop from whichever projector is on, to the next one in sequence, A–B–C–A– etc. In this case, it is assumed that the dissolve loop, like the projectors, is set up in groups of three. However, this is not a necessary condition, and some programming systems permit the setting up of different dissolve loops within the same system. For example, a fast-moving show might use three center screens, each with six projectors working on a six-projector loop, and two side screens, each with three projectors on three-projector loops. Systems that allow these types of dissolve loop setting usually allow the setting of loops of two, three, four, five or six projectors.

Cues 002 and 003 also include a T command. This is an automatic 'wait' or 'time to next cue' command, which means that when the show is actually run cue 003 will automatically be released at a set time after cue 002, and that cue 004 will likewise follow cue 003 automatically. T1.5 means the time to the next cue is 1.5 seconds. Again, it is possible to select times to any time up to, say, 10 seconds. Most systems allow at least ten cues per second, and the best allow at least twenty per second, so in this example a command T0.05 would be valid. Such close spacings are used for ripple and zoom effects.

Cue 006 is an animation cue. In the particular system used as an example, each projector lamp can be instructed to flash at a wide range of different rates, from ten times a second to once every ten seconds. The ratio of on to off times can also be selected. A510 could mean flash on and off once a second with an equal on and off time. In the example projector 2C is so instructed at the time its lamp is off, so cue 007 brings it on, in the example quite quickly, but it could be faded on as a flash fade.

Cue 10 introduces the idea of a loop of cues. The command L04 means: 'Carry out the following cue sequence four times'. In the example the idea is to ripple a change across all three screens four times.

The ripple is quite fast because the time between cues within the loop is only 0.3 seconds. The command D0! signifies that the projectors must not step. Usually when a projector lamp fades down, the slide in that projector automatically advances, because this is what is usually wanted. However, in some animation and superimposition sequences the same slide is wanted again, so a 'do not step' command is needed. In the system shown as an example, any command or individual projector required not to step is identified by '!'. By the way, in the example D0 means dissolve at zero seconds, i.e. move as quickly as possible.

Cue 014 identifies the end of the loop by LE (Loop End). Loops can be up to 100 cues long, and there can even be loops within loops. Cue 015 is an example of a cue being used for a text description. This cue does not actually do anything other than remind the programmer of where he is in the program.

The task of multi-image programming is made easier by the fact that all the projectors may be connected up while programming is in progress. As each command is entered, the projectors execute that command, so it is easy to see that the correct slides are being shown. At any time the operator can go to any selected cue, and all the projectors will back up or advance to their correct position corresponding to the selected cue. A program can be run in whole or in part, with the cues released manually by the operator, or automatically wherever there is a T command present.

Synchronizing to sound

There are some applications of multi-image which only require the program to be put into the computer, and are then manually released using a remote cue button or the computer's space bar. Examples include speaker support sequences in sales conferences, and the use of multi-image in the theater. Often these applications will have automatic sequences where a chain of cues use the automatic cue-release feature, and the computer comes to rest at the end of a sequence to await the release of a subsequent single cue or chain of cues.

For these applications the program should have some special staging features, such as the ability to black out all the projectors in an emergency, and to be able to step backwards through the cue sequence.

Most users of the medium, however, want their multi-image sequence synchronized to a sound track. While it is possible to transfer the show data

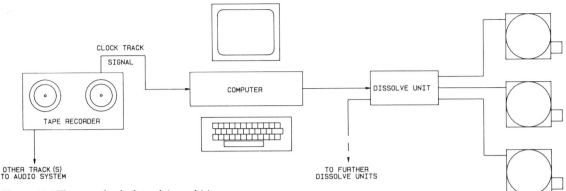

Figure 16.4 The use of a clock track in multi-image production.

direct to magnetic tape to run in synchronization to the sound (simply by manually releasing the cues while listening to the sound track), this method is neither elegant nor precise. The preferred technique is to use a clock track.

Here one track of the sound tape is dedicated to a clock signal. The nature of this signal is such that the time is recorded in a unique code as minutes, seconds and fractions of a second. Thus, the show producer can identify a spoken word or beat of music by its clock time, opening up the way for absolute synchronization between the sound tape and the computer.

The usual procedure is to use a combination of 'clock cues' and 'time to next cue' commands. Thus, pre-timed visual effects will use the equivalent of the T command, but the initiation of such a sequence, or of individual widely spaced cues, will be by providing the cue with a specific clock time.

The show is reviewed by playing the clock track into the computer, which then releases the cues at the required times. If the timing of individual cues, or even the whole show, looks wrong, simple editing commands can adjust individual times, or make offsets to groups of cues or the whole show.

Clock tracks are of two kinds. The manufacturers of multi-image equipment may have their own proprietary code. Generally, this results in a lower-cost system with a code that is easy to copy, and which may typically resolve to 0.1 second. The computer can interpolate commands where a 0.05 second cue spacing is required. These codes are entirely satisfactory for those making pure multi-image shows.

Recently, there has been a great tendency to make video programs by multi-image methods. The show is first made as a single-screen multi-image show, using anything up to eighteen projectors, and then transferred to video using an optical multiplexer. If it will be necessary to mix in material with conventional video, it makes the editor's task much easier if the computer uses the SMPTE or EBU clock code.

Clock codes were introduced into video for the same reasons that they were introduced into multi-image. They make the linking of different sources easier and allow accuracy in synchronization. The SMPTE code works in hours, minutes, seconds and frames (30 frames per second for video). The EBU code is the European equivalent at 25 frames per second. There is also a motion-picture variant at 24 frames per second. The only disadvantage from the multi-image producer's point of view is that computers fitted with the SMPTE/EBU code facility are a bit more expensive than those using a house code, and they may prefer to work in easy fractions of seconds rather than in notional frames.

Control tracks

Some users keep the computer on-line for running the actual show, especially where mixed live and recorded sequences must be shown in one presentation. Many users, once they have seen and approved the original programming, will want to have their show transferred to magnetic tape, so the one tape carries the sound track and the control data. This makes it possible to use simple replay-only equipment to run a show, and eliminates the need to cart around the computer. Thus, show copies consist of a copy of the audio directly from the master tape, but have a control track automatically laid down by the computer, in precise synchronization resulting from its reading of the clock track on the master tape.

Table 16.2 Characteristics of some commonly used multi-image control signals

Manufacturer	Signal name	Number of projectors	Cue spacing	Bandwidth	Aux A	Aux B	Signal
Clear Light	Cue Sentry 20	15	0.05 s	4 kHz	5	33	SFBP
Clear Light	Cue Sentry 50	15	0.02 s	8 kHz	5	33	SFBP
AVL	Positrak	15	0.05 s	10 kHz	10	25	FSK
Arion	Mate Trac	16	0.05 s	2 kHz	12	12	FSK
Electrosonic	Alphasync	24	0.05 s	6 kHz	24	96	FSK

Aux A	Number of auxiliaries available from dissolve units.
Aux B	Number of auxiliaries available with extra equipment.
Bandwidth	Recommended −3 dB response of tape recorder.
SFBP	Single frequency, Bi-Phase.
FSK	Frequency shift keying
Data	Data on the tape in all cases includes slide position information as well as the projection effect instructions.

The nature of the control track recorded on the tape is extremely complex. In all professional systems the control track carries not only the main programming commands, but also a continuous update of slide position information. This means that, in principle, if a tape is started half-way through, all the projectors move as quickly as possible to the correct position. Unfortunately, due to the slightly different capabilities of the different systems on the market, each manufacturer has their own control signal. This is not as great a disadvantage as it may sound, because multi-image is a medium that is applied to single events and single venues. Some manufacturers offer limited compatibility for the replay of shows made on other systems.

A track on standard audio tape is limited in the amount of information that it can reliably carry. There is a trade-off between the number of devices that can be controlled and the speed of cue release. Thus, in theory, a system that can control fifteen projectors at twenty cues per second could be offered in an alternative version that controlled thirty projectors at ten cues per second. In practice it is easier to limit the system to one defined capability, and some examples are shown in Table 16.2. All professional systems include a complex arrangement of error correcting to ensure that the data is recovered uncorrupted; even so, it is best to make show copies as direct first-generation output from the computer, rather than as a tape-to-tape dub.

Big shows

Capacities of some standard multi-image computers are shown in Table 16.3 If bank switching is necessary, this means that the available cue rate is correspondingly reduced, and it is not possible to address all projectors simultaneously. (They must be addressed in banks of, typically, thirty projectors.) For most users of multi-image, anything above twenty-four projectors is fairly academic, but even at this level there can be a problem.

Table 16.2 shows that any of the common tape-control signals can only control a maximum of twenty-four projectors. This means that a large playback system may need to use more than one control track. Because modern multi-track tape recorders are such good value, this is not a serious problem. A 72-projector playback system using three control tracks is a perfectly reasonable proposition. However, above three control tracks there would be a tendency to use one or more

Figure 16.5 A big show multi-media computer. This one can handle 192 slide projectors, 256 dimmers, 1536 auxiliaries and many video devices – all at 30 cues per second. One VDU is used as a cue display, and the other as a status display.

Table 16.3 Characteristics of some commonly used multi-image computer systems

Manufacturer	Computer	Language	Projectors A	Projectors B	Cue spacing	Clock
Clear light	Apple IIE	AMPL+	240	30	0.01 s	P
AVL	AVL Genesis or	PROCALL 5 or	120	30	0.05 s	P or
	IBM PC	PROCALL X				SMPTE
Arion	IBM PC512	Arion	512	32	0.01 s	P
Electrosonic	Apple IIE	ESCLAMP 24	24	24	0.05 s	P
Electrosonic	Apple IIE	ESCLAMP 48	48	24	0.05 s	P
Electrosonic	IBM PC	MSC-48	48	48	0.033 s	SMPTE+
Electrosonic	ES MSC	MSC-192	192	192	0.033 s	SMPTE+

P	Proprietary clock code
SMPTE	SMPTE clock code
SMPTE+	SMPTE and EBU clock code. 24, 25 and 30 cues per second.
Projectors A	Maximum number of projectors programmed.
Projectors B	Maximum number of projectors that can be addressed on one cue.
Computers	In all cases where an Apple IIE or IBM computer is used it is also necessary to add a communications and/or clock card or similar device to complete the system.

computers on-line with only one clock track. In such cases it is possible to commit the show to a solid-state memory, for example Bubble Memory or EPROM, for greater security.

Most of the multi-image systems can control devices other than slide projectors. These devices are usually referred to as auxiliaries, and are available both at the projector interface, for controlling effects local to the projector, or, more conveniently for large systems, at separate auxiliary interface units. The 'big show' computers also allow access to other devices which require analog control, such as dimmers and servo-motors, and to devices which require complex addressing, such as videodisks.

'Big show' computers reveal a further problem. In a large planetarium or major theme park entertainment there may be several hundred devices under the control of one computer. In small multi-image systems it is easy to display the status of all the devices being controlled, either on request or on the same screen that carries the cue information. In larger systems it becomes helpful not only to have an extra monitor for displaying status, but to have a computer program that allows status information to be laid out to match the actual information. The program should also allow controlled devices or groups of devices to be identified by names rather than anonymous numbers.

The top end of the multi-image market merges completely with multi-media, where slide projection becomes just a part of the complete show package. Increasingly, it will merge with other media for special events or one-of-a-kind installations. Motion picture is easily integrated with multi-image, particularly if time code techniques are used. So is video, but this opens up another intriguing possibility. Multi-screen video presentations using programmable frame stores will find a useful place in the armoury of spectacular display. The people who are familiar with multi-image techniques will have no difficulty in programming them on the same computers that program multi-image, using a suitably revised program.

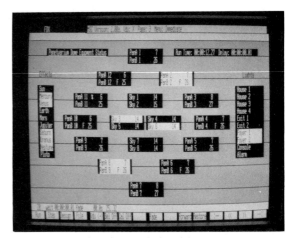

Figure 16.6 For big shows it helps if the status screen can be laid out in a way that matches the controlled equipment's location and function.

Chapter 17

Movie presentation systems

The movie film is the grandaddy of the AV business. Although it could be argued that the magic lanternist with his patter and well-rehearsed routines was the first AV practitioner to use projection, it was the sound film that, at one bound, became the ultimate in audio-visual. Strange, though, that the medium should be so perfect and yet at the same time so limiting.

Perfect, because if a moving-image sequence with sound is required, movie still remains the ultimate method of high-quality program production; limiting, because once made, a film is difficult to adapt, to keep up to date, and to use for different purposes.

Movie film is still the most important method of high-quality program production, wherever the program involves movement and is destined for a wide audience. The industrial AV user's interest may be more limited, but nonetheless movie has an important role to play. This chapter is mainly concerned with the various mechanical possibilities of presenting movie film, but in reviewing these it is possible to define the uses of the various movie formats. The comments here are from the point of view of the AV user and do not take into account the amateur market.

Super 8

This is not a serious production medium. It is now also losing to compact video presentation units as a showing medium. However, the compact portable desktop Super 8 unit with built-in rear projection screen is smaller, lighter, less expensive, and gives a bigger picture than its all-electronic counterpart. These types of systems use Super 8 in endless cartridges to simplify operation.

16 mm

This medium is the most important for the AV user. It is the medium in which sponsored films are made, and is very often the preferred production medium even when the program is intended to be finally shown on video. It is the nearest thing there is to a world audio-visual standard. If it is necessary to guarantee that a program can be shown anywhere in the world, then 16 mm is best because a 16 mm projector can be found anywhere and, in contrast to video, standards are uniform.

Sponsored films are usually made by those organizations which need programs for a distribution. Thus, shows intended for distribution to schools, training films (especially sales, services

Figure 17.1 Portable unit for individual presentations using Super 8 film in a plug-in cartridge. (Photo Edric Audio Visual Ltd)

and management training films) are best on 16 mm, even if the same films are also available on videocassette for small audiences.

If a show has been originated on 16 mm, then for all but the very small audience, it should be shown on 16 mm. The impact is significantly greater, particularly because making the effort ensures a better environment for the show.

Most 16 mm presentations are given using portable 16 mm projectors. Present-day portables are lightweight and easy to use. They are available in self-threading versions, which make them no more dificult to use than a cassette tape recorder. Like slide projectors, they are flexible in positioning because a range of lenses is available. Fixed focal length lenses are best, but zooms are available.

The open-gate light output of a standard 16 mm projector with the shutter running is about 700 lumens. This makes it suitable for pictures up to 4 m (13 ft) wide in the dark, especially if a screen of reasonable gain is used. (This may not be possible; see Chapter 22.) If bigger images are required,

Figure 17.2 Modern 16 mm projectors are lightweight, compact and portable. Self-threading models are particularly easy to use. (Photo Elf Audio-Visual)

then instead of the tungsten halogen lamp used in the standard portables, it is necessary to move up to a projector fitted with a xenon arc light source. Portable xenon arc projectors, using 300 W or 500 W lamps, are available with light outputs of up to 2000 lumens suitable for screens of 5–6 m (16–20 ft) width.

For permanent installations and other applications requiring big images, pedestal-mounted projectors with 1000 W and 2000 W xenon lamps are available, and these can give pictures in widths of 8–10 m (26–32 ft). These projectors are usually fitted with big reels that allow the uninterrupted showing of feature-length films.

35 mm

This is the professional standard which is used for the making and showing of entertainment films. Industrial AV users will not usually come across it, unless they are users of TV commercials, some of which are originated in 35 mm; or because they have the need for a spectacular show (e.g. a product launch) where 35 mm may be essential in order to give adequate picture quality on a big image.

There are 35 mm portable projectors available, but these are usually used for preview work. They do not have a significantly greater light output than a 16 mm projector. Most 35 mm projectors, however, are in permanent installations. They are heavy-duty pedestal-mounted machines intended for many years of continuous service.

Because of the varying requirements of different installations, 35 mm projectors are of a modular construction. For example, there is a choice of lamp houses using 1 kW, 1.6 kW and 5 kW xenon lamps to serve the needs of screens of 6–17 m (20–56 ft) width. There is also a choice of sound system and film handling arrangements.

The traditional movie theater uses two projectors. The film comes on a number of reels, each only holding about twenty minutes worth of film, and the projectors change over at each reel break. A simple system of interlocked shutters ensures that the audience are unaware of the changeover.

Most modern movie theaters, however, run their programs on a single-projector basis. This requires special film handling equipment, because a reel to carry a two and half hour program is enormous. Two methods are used. One is the use of big reels, vertically mounted, either on the projector or on a separate stand. Special arrangements are then included for the rapid and safe rewind of the film at the end of the show.

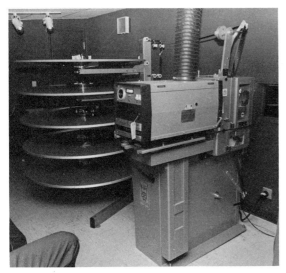

Figure 17.3 A typical 35 mm movie projector installation. Here a five-platter non-rewind system has been installed to allow easy changes of program. (Photo Electrosonic Systems Inc.)

A more elegant method, which requires more space, is the use of the non-rewind platter. Here the film is stored on a large horizontal plate or platter. The film is drawn from the center, and is then fed back to a platter on another level. This, of course, fills up from the center, so when the show is finished the next show can be rethreaded from the new platter without the need for rewind. Most installations use a three-platter system, with two platters in use for the show and the third one available to make up new programs, or to strip down old ones.

Most modern movie theaters are at least partially automated. This means that the film itself is used to initiate the operation of drapes, lighting dimmers etc. The usual method is to apply a small piece of flexible metal foil to the film at the required point. This is then detected by a proximity detector or infra-red photocell.

70 mm

Unless it is necessary to go into the public entertainment business 70 mm is of even less relevance to the industrial AV user than 35 mm! However, it is the key to spectacular big-image projection.

It was originally introduced as an entertainment film format for big pictures, and the film was both shot and shown in wide gauge. In recent years it has become a showing medium only, for normal feature films. This is because the quality of the negative films used for shooting the original film has improved so greatly that today's 35 mm negative film can give as good resolution as yesterday's 70 mm. However, when it comes to making the show copy, better presentation quality is obtained on 70 mm, so the blockbuster feature films are often transferred to 70 mm for use in prestige movie theaters.

A number of special-effect shows in theme parks and other tourist entertainments use 70 mm. These sometimes employ lenses of very short focal length to give the audience the impression of being in the picture. In these shows it is important to use projectors with exceptionally good picture steadiness so that viewing is not uncomfortable.

Without doubt, the most spectacular 70 mm films to be seen are those on the IMAX process. In this Canadian system, films are made by running 70 mm film horizontally, and the actual show uses a projector working on the same principle. This

Figure 17.4 The well-known Simplex projector mechanism is available in a form which accepts both 35 mm and 70 mm film. (Photo Strong Inc.)

ensures an original that is far bigger than that achieved in 35 mm or even normal 70 mm production, resulting in exceptional quality. The showing system is one where the audience are quite close to the screen, so the picture must be rock-steady. This is achieved by pin-registering each frame as it is projected, and using a rolling-loop system of film transport to prevent tearing of the film.

IMAX requires a special auditorium. The screen may be as big as 30 m × 22 m (98 ft × 72 ft). The sound is run from a separate interlocked sound system. IMAX is only suitable for short subject films. It has a near relation, OMNIMAX, which uses the same type of projector with a fisheye lens to fill a whole hemispherical dome, usually a planetarium dome that has been tilted. OMNIMAX is a suitable support for mixed-media shows, being given in a dome, but is a less satisfactory medium than IMAX. IMAX gives huge pictures of superb quality, with excellent sharpness, contrast and saturation. OMNIMAX suffers from the use of a dome as a screen, where cross reflection reduces both contrast and saturation, and where it is difficult to get a lens that will maintain perfect focus over the entire picture.

Another special-purpose 70 mm movie format is Showscan™. This was developed by Douglas Trumbull in Los Angeles as a medium for spectacular shows to be used in theme parts, simulator rides and Expos. There are two prob-

lems associated with normal movie film, which the Showscan process seeks to overcome.

The first is that there is a limit to how brightly a picture can be projected. This may seem odd, because in many projection applications the problem can be getting enough light. However, in the dark, a picture that is too bright suffers from objectionable flicker as a result of a breakdown in the viewer's 'persistence of vision' mechanism, which normally allows him to see the 24 frames per second as continuous movement.

The second problem is that the relatively slow picture rate of 24 frames per second means that, if there is a very rapid movement, the jumps between individual images are large, so the movement has a blurred quality. In addition, stroboscopic effects are badly noticeable on rotating wheels etc.

The Showscan solution is to run standard 70 mm film at 60 frames per second. This allows a much brighter image to be projected without flicker, reduces stroboscopic effects, and makes rapid movements much sharper. The result is a much more life-like picture.

Movie formats

Although the gauges of film are standardized, there are variations in the format of picture that is

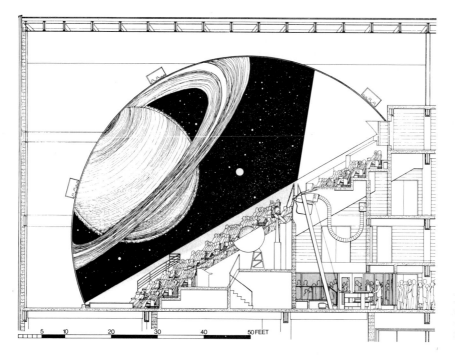

Figure 17.5 OMNIMAX is shown in a tilted dome. (Diagram courtesy Imax Systems Corporation)

Figure 17.6 High-quality sound in the movie theater is achieved using a Dolby™ cinema sound system. This processor is used for showing films with stereo optical-sound tracks. (Photo Dolby Laboratories Inc.)

shown. Originally all films were made in the Academy ratio, which is 4:3. This is still the format of television pictures.

Most industrially sponsored films, and all films made for video, are still originated and shown in Academy ratio, so AV users usually have to contend with this format, which is different from that of slides (3:2). Entertainment films, however, are now normally presented in a wide-screen format. This means that when making the film the director has to compose his shots to make reasonable sense both in wide screen and in the truncated Academy format. In fact, when films are shown on TV the major broadcasters have a system of automatic panning to ensure a sensible picture composition on the narrower screen.

The simplest method of achieving wide screen is to use a mask in the projector gate with the different aspect ratio. In the commercial movie theater films are shown with an aspect ratio of about 1.85:1. To get a wider screen an anamorphic lens is used to 'squeeze' a wider picture on to the film during photography and to 'unsqueeze' it on projection. This method makes better use of the film area and light available. The final projected ratio is between 2.2 and 2.5:1.

Movie sound systems

The great majority of 16 mm and 35 mm films are shown with a mono optical sound track. This is the most reliable and simple form of synchronized sound, and it is recommended for all standard film presentations. Sound is recorded photographically on the edge of the film as a variable-area optical track; the area of exposed film at any moment is an analogue of the momentary signal value. The optical track is scanned by a photocell.

The frequency response of optical sound is limited. However, on 35 mm film it is now possible

to have Dolby™ optical sound. This uses a process similar to that used in the Dolby noise reduction system for magnetic tape to increase the frequency response and reduce the noise of optical sound tracks. The results, in practice, are as good, if not better, than magnetic tracks.

When there is a need to be able to change the sound track on an existing film, or to have extra tracks, magnetic sound is used. Most portable 16 mm projectors have magnetic sound as an option.

Multi-track sound in the movie theater can be achieved by multiple magnetic tracks carried on both sides of the 35 mm film. Typically there are three main tracks and an effects track. However, this format is now not used very much. Movie theaters either use Dolby stereo optical sound, or, if sound is to be a major feature of a big show, use multi-track magnetic on 70 mm film, also with Dolby noise reduction.

When film sound tracks are made, the sound starts off on a separate magnetic medium, and is only dubbed on to the show copy as the last stage in production. During production, the sound track must be made and edited in frame sync with the picture. This is usually done by recording the sound on to magnetic film which is sprocketed in the same way as the picture film.

So-called double head projectors allow the running of SEPMAG sound, where the sound track is carried on a separate reel of magnetic film. Because the projector motor drives both reels together, they stay in exact sync.

During production, however, the sound is usually on a separate machine. The interlocking of the two machines is achieved by electromechanical means (Selsyn interlock). For special shows requiring high-quality multi-track sound the same system may be used on the final showing. It is possible to interlock projectors using the same technique, so 'multi-screen movie' using from two to as many as nine 35 mm projectors with a separate sound machine is quite feasible.

Recent electronic developments allow the frame synchronization of ordinary magnetic tape with movie projectors. This requires either a simple pilot tone, or more usually a clock code on a separate track of the magnetic tape. Although it is most useful in ensuring high-quality sound track production, the method can also be used in multi-media showing systems.

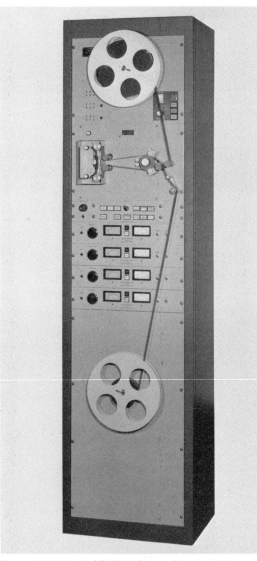

Figure 17.7 A typical SEPMAG sound transport for 35 mm magnetic sound. (Photo Magna Tech Inc.)

Mixing movie with multi-image

In produce launch shows and some public entertainment shows it is often necessary to be able to mix movie with multi-image slide. This is quite easily done although, in the case of permanent installations, the intending user should carefully examine the likely impact on running costs.

Slides are capable of giving superb photographic quality, and one of the problems of mixing movie with slides is that the movie can look disappointing. To avoid this it is essential that a 16 mm film is not required to give an image bigger than the corresponding slide. For example, if the format is a screen 6 m × 2 m (20 ft × 6.5 ft) with effectively two slide images side by side, the biggest permissible image from 16 mm is the Academy ratio picture 2 m high, i.e. 2.66 m × 2 m (8.7 ft × 6.5 ft). If a bigger picture is required, it is *essential* to use 35 mm. In this case, by using anamorphic projection with a special aperture plate, the full 3:1 format can be achieved.

The best method of synchronization depends on the nature of the show. The rule is to use the *simplest* method that meets the show requirements.

Most shows require inserts of movie. Here, it is best to program the start(s) of the movie projector from the programming system, and to fit the movie projector with an automatic stop system. To ensure that the image appears on the screen at exactly the right moment, and also leaves the screen on cue, the light output of the projector is also programmed. This is done with an automatic dimmer. In the case of small projectors this is achieved with tungsten halogen lamps. In the case of xenon arc lamps a solenoid-operated douser (shutter) is used.

Usually it will be quite sufficient for the sound track to continue on the separate magnetic tape. Synchronization will be quite good enough for any program with music and commentary. However, if a short piece of lip sync is required, the programming system can switch in the movie projector's sound track for the necessary duration. This method of linking movie and slides is reliable, technically simple and easiest for production. It meets most requirements.

There are some shows where extended lip sync is required, or where the movie element is not just a few small sections of the show, but is effectively running right through the show. Then there are several alternative possibilities:

☐ To run the whole show from the film, using multi-track magnetic on the 35 mm picture film. The

Figure 17.8 The projection room at The London Experience. This show uses 30 slide projectors, multi-track sound and 35 mm movie carried on an endless loop.

so-called effects track is used to carry the multi-image control data, leaving three sound tracks. Technically this is a good method, but these days it is difficult to find studios that can do the dubbing. It is probably now only applicable to 70 mm.

☐ To use a separate film transport electromagnetically interlocked with the movie projector. These 35 mm magnetic film transports can give up to six or eight tracks.

☐ To use an electronic method of synchronizing a reel-to-reel tape deck with the projector. Several are available. The most sophisticated use a clock code on both the film and the tape to ensure absolute synchronization; others rely on a correct start-up procedure. All require that small adjustments can be made to the running speed of either the projector or the tape deck during the show. Although it is easier to adjust the speed of the tape deck, it is better to adjust the projector speed.

Continuous-running films for exhibitions

Sometimes it is necessary to show films on a continuous basis, particularly in exhibitions and permanent displays. Shows requiring only a small screen are now often presented on video, on the grounds of lower running costs and easier maintenance.

However, where the enhanced picture brightness and quality of real film are needed (especially, for example, in mixed-media shows), automatic film showing is better. With 16 mm film there are various possibilities;

☐ Projectors for 16 mm are available that automatically rewind the film after each showing. This system is

suitable for small theater-style shows. It avoids the problem of film rubbing against itself.

☐ Endless loop attachments are available for 16 mm film. These should only be loaded with Mylar-based film that has been specially coated for use in endless-loop applications. These attachments are compact.

Figure 17.9 A 16 mm endless-loop attachment. (Photo Bergen Expo Systems Inc.)

Figure 17.10 Close-up of a 35 mm endless-loop film handler. (Photo Electrosonic Systems Inc.)

☐ Another form of endless loop, suitable for short programs, has the film run on a big open rack round a lot of pulleys. This system avoids the problem of film surfaces rubbing against each other, but uses a lot of space.

☐ Any system of continuous film projection should only be operated in a clean atmosphere. Some totally enclosed systems are available, of both the compact endless loop and open rack variety, that are built into enclosed cabinets which have their own air filtering, continuous film cleaning and controlled humidity arrangements.

The principles are the same for the larger film gauges. Short shows use rack-type film loops. It is possible to get horizontal endless loops for 35 mm film with very long playing times. These loops carry the film in such a way that film surfaces do not rub against each other. They need a minimum of 40 minutes worth of film to operate correctly, so for some multi-media entertainments it is necessary to load two or more copies of the show onto the loop. However, many applications involving 35 mm projection with relatively long shows can tolerate a rethreading interval every few hours; and these applications can use the standard platter equipment.

Chapter 18

Video production

Most industrial and business users of video are concerned with the *showing* of video programs. Thus, Chapter 19 which discusses video presentation systems is likely to be of most interest. This chapter covers a few topics on video production which may be of relevance to the person commissioning video programs, or to someone seriously considering establishing an in-house production facility.

This chapter is *not* intended as a short primer on how to make a video program. Anyone who needs detailed information of this kind should consult appropriate text books.

In principle, the making of a video program is no different from the making of any other AV program. The processes described in Chapter 4 must be gone through in the usual way. The differences with video can be summed up as:

☐ The enormous variation in cost between low-quality equipment and professional equipment.

☐ The immediacy of the medium, with its ability to replay material as soon as it has been shot.

☐ The electronic nature of image storage which allows special effects to be easily achieved by processing electrical signals.

The very high cost of broadcast-standard video equipment has led to a situation where nearly all video post-production (editing and audio mixing) is done at facilities houses. Few program production companies have their own broadcast-standard production facilities, so they will usually plan the work they need doing, and book the necessary time at a suitably equipped editing suite. Because editing suites may well cost $400 per hour, the planning is all-important if budgets are not to over-run.

There are various levels at which video programs intended for business and industrial use can be made. The choice of which level is appropriate must be made on the basis of the intended application. The levels are:

☐ **Broadcast standard.** This is the highest standard used by professional broadcasters. This standard would usually also be adopted by those making TV commercials for public showing, and by those making prestige programs for corporate image use. It would also be appropriate to any program that will be required in a large number of copies.

☐ **High intermediate.** This standard would only be just below broadcast. It would be used for programs for which wide distribution was expected. The main difference between this and broadcast standard is that the cameras would not be so expensive, and the range of special effects available would be more limited.

☐ **Low intermediate.** This is the standard appropriate to the majority of business, industrial and educational users. The aim is a good commercial quality suitable for copying.

☐ **Training.** This is a polite word for low standard. Here, the equipment used is of good domestic standard. The results are fine for role-playing training applications, experimental work, and limited-interest programs where distribution is not required.

All four levels can be relevant to the commercial user. But because the equipment prices for complete production systems range from $7,000 to $1,000,000 it is important to decide which level is required.

Videotape recorders

The type of videotape recorder used to make the program is a good guide to the production level aspired to. Unlike movie film, where there are only two types of film, 16 mm and 35 mm, relevant to commercial production, there are at least four relevant tape standards for videotape, with more looming.

153

Until recently, broadcast studios used tape machines with 2-inch tape, but now this has been largely replaced by 1-inch tape. The helical scan 1-inch broadcast machine is the mainstay of professional video production. Apart from the ability to record a high-quality image, to be able to run at different speeds and to give a good still-frame performance, the feature that distinguishes the broadcast machine from the rest of the pack is the ability to maintain quality through several generations of copying.

The process of video editing is completely dependent on copying signals from one tape onto another. It can thus happen that a show copy of a video program is at least two generations, and often three or four, away from the master material. It is the broadcast machine's ability to maintain quality through several generations that sets it apart.

There are compact portable 1-inch machines available for location work, but these are a comparatively recent development. The requirement for something more portable than the 1-inch machine for applications such as news gathering led to the introduction of the high-band three-quarter-inch videotape recorder. This uses tape in the U-Format cassette (see Chapter 19) and gives an initial picture performance similar to that of a 1-inch machine. Although it is accepted for broadcast purposes, it is not quite as good at maintaining quality through the generations.

Recently high-band U-Format has been joined by the arrival of the broadcast-standard 'Camcorder' for news gathering. This is a compact assembly combining a high-quality camera with a compact video recorder. The main manufacturer is Sony, with their Betacam™ machine. Although this uses a cassette with half-inch tape that is similar to that used in their domestic machines, there is no other similarity. The signal is recorded in an entirely different way and is able to give images of superb quality.

The high intermediate standard is based on the use of high band equipment. The words high band refer to the video bandwidth recorded, which is itself a measure of the picture quality and detail.

The low intermediate standard is, not surprisingly, served by low band equipment. This also uses three-quarter-inch tape in the U-Format cassette, but cannot give the same quality as its broadcast equivalent. It is, after all, only a quarter of the price! However, low band U-Format is a robust medium and is the basis of most industrial video.

At the bottom of the pile, systems based on half-inch tape are available. By far the most popular are complete systems based on the VHS tape format. The main problem with VHS is that it is not possible to maintain quality when copying through several generations. It may, therefore, be quite suitable for use in a training department, where program copies are not required, but is not recommended for serious production work.

It is important to understand that the way in which programs are made need have no influence on how they are shown. Usually, shows made on high band are later copied over to low band or VHS for final showing. The rule is to use as high a quality as one can afford to originate the program, and then make copies to match the showing facilities.

Cameras

Video cameras come in as wide a price range as video recorders, so usually the two are matched: broadcast cameras with broadcast recorders, domestic cameras with VHS. The aim should be to use as high a quality camera as can be afforded.

To record a color picture, it is necessary to split the picture information into its red, green and blue components. Broadcast and professional cameras are usually three-tube cameras. The incoming light is split into its three color components using dichroic color filters, and each component is fed to its corresponding pick-up tube. Although this system gives by far the best color quality, there is a problem of registration, i.e. ensuring that in the

Figure 18.1 The JVC KY 2700E camera is a popular high-quality industrial video camera. (Photo Bell and Howell AV Ltd)

resulting final image all three components correctly overlap. In the old days the lining up of a broadcast camera could take an hour or more, the latest cameras include automatic registration.

All low-cost cameras, and even some professional ones, are single tube cameras. Here a single pick-up tube has a three-color grid filter, and the three color signals are extracted electronically. The problem is that the filter necessarily limits the definition; close examination of the resulting image can show a stepped pattern to any color edge.

Camera performance can be measured in terms of the objective quality of the image and in the facilities offered by the camera, e.g. by the type of lens fitted. Apart from the obvious items like lack of picture distortion, picture sharpness, good color rendition, there are three special attributes that distinguish a good camera from an also-ran:

☐ The ability to perform well in low light conditions without introducing graininess.

☐ The absence of any picture smear when the camera is rapidly tracked. Similarly the minimization of any 'comet tail' effect arising from the tracking of a highlight.

☐ The ability to be precisely set up to match a standard, and to maintain the set-up without further adjustment.

The last point is particularly important in studio applications. Most studios will use at least three cameras in order to provide a sufficient variety of shots to give a fluent production. It is clearly vital that the results from each camera are identical. It would be a disaster if, in an interview, the flesh tones of the participants changed as different cameras were used.

Vision mixers

It will already be clear that a video production studio is a considerable investment. This is why only those who are reasonably sure that they can make full-time use of such a facility should consider making the investment. In addition to the videotape recorders, studio lighting, audio recording facilities, cameras and all their related accessories, the assembly of a video program in a studio will certainly require the use of a vision mixer.

This device can select from a number of different video inputs, usually the three cameras, plus all sorts of other picture sources such as other videotape machines, caption generators etc. The person using the mixer can usually see all the material available on corresponding preview monitors, and will then select the required image either by switching or by cross-fading. Modern mixers offer a variety of transitional effects including many kinds of picture wipe.

When all this equipment is hooked together, unless all the picture sources are in sync (i.e. the timing of their scanning of the video image is precisely in unison) there are nasty picture rolls as each transition is made. Thus, within a studio, all equipment will work to a 'master sync' pulse that is distributed to all the equipment being used. The timing of video signals is sufficiently critical that adjustments even have to be made to take into account different cable lengths.

A particular problem arises when it is necessary to cut to a signal derived from another videotape source. In this case, it is difficult to ensure that the source recorder is perfectly in sync with the rest of

Figure 18.2 A vision mixer. (Photo Michael Cox Electronics Ltd)

Figure 18.3 The Aston 4 character generator. The keyboard is what the operator sees, but behind it is a considerable array of electronics. The system allows a huge choice of fonts, character sizes and coloring.

the system, and even if, in principle, it has been locked to the master sync, there will usually be a residual loss of definition. This is put right by the use of a timebase corrector. This expensive piece of equipment is an essential part of the studio armory, and is often used to clean up material that would otherwise be sub-standard by the time it reached the master tape. Low band original material is enhanced by its use, so if this type of material must be used as part of a professional production, it can be suitably sharpened up to match the new material.

Many other devices are now considered to be an essential part of the video production studio. One is the electronic caption generator. Captions used to be put up on the screen by pointing a camera at a suitable piece of artwork, or at a roller caption machine. Now digital electronic caption generators allow the generation of captions in a wide choice of fonts and in any desired color combination.

The digital processing of video images opens up a whole range of possibilities for the manipulation of image sequences. Now that startling video effects can be seen every day on television, it is not surprising that industrial video users are asking to have similar effects within their shows. The best special-effects generators are still extremely expensive, so most users rent by the hour, and, most importantly, obtain the services of someone who knows how to operate the system. However, relatively low-cost systems, with a corresponding reduction in performance, are becoming available for the semi-professional user.

One trick that is provided on even relatively low-cost vision mixing systems is color separation overlay or chromakey. The principles have been known for years, and have formed the basis of many special effects in the movies. Electronics just make it easier. The process is used when a studio shot must be married to some other, usually exterior, shot. For example, if a studio commentator wishes to appear to be speaking from some exotic location.

The usual arrangement is for the presenter to stand against a blue background, either a white cyclorama illuminated with blue lighting, or a blue cyclorama illuminated with white lighting. The chromakey system is then fed with two inputs, the studio signal and the remote signal – for example, a color slide of the south of France or a film of a rocket launch. Whenever the chromakey device detects a saturated blue signal, which is easy for it to detect, it substitutes the alternative signal. Thus, the final image is a composite of the two images.

Two problems arise with chromakey. The first is the obvious one that the foreground shot must not include any saturated blue. A blue shirt will result in an interesting south of France pattern. The second is that hair and other fuzzy outlines cause the system to dither and to give the foreground a halo or cardboard cut-out quality. The more expensive the system, the less the problem.

Video editing

The editing of videotapes is essentially a copying process. The principles are easily understood, and anyone can learn how to press the appropriate buttons. However, video editing is what puts the polish on a video production, and good editing can make all the difference between an excellent and a mediocre show. Creative video editors are a rare breed, and at the top end of the market the fortunes of an editing facility with millions of dollars invested in equipment can depend on attracting and keeping this talent.

A simple editing suite consists of two machines, the source machine carrying the original material, and the master machine onto which the master tape is being assembled. The simplest method of editing is known as assembly editing and involves sequentially dubbing the required sections of the original tapes onto the master machine.

If the material was simply copied across there would be a bad picture roll at each transition, owing to the lack of synchronization between the machines. The two are therefore linked by an edit controller. This allows the editor to select the required section, and then command an automatic edit. This process involves a pre-roll, where each machine backs up (usually five or ten seconds), and then comes forward using the pre-roll time to get into precise sync. At the designated frame the master machine switches from replay into record and records the required section. Often stopping is automatic because the editor can also designate the stop frame.

Professional video people found a lack of precision in this simple method, particularly in the rapid location of material, and in being sure that particular sequences were going to fit their allotted slot. Professional editing now tends to be done using the insert editing technique, based on the use of time codes.

Here all material is striped with a unique time code. This is either 'linear time code' carried on a spare track of the tape like an audio track, or 'vertical interval time code' carried within a spare part of the picture signal. Its importance is that it allows any required section of source material to be located rapidly, and moved to any required place on the master tape. Even a single frame can be placed at a designated frame time.

Not surprisingly, computer techniques are essential to this style of editing. The computerized edit controller will not only precisely control up to four videotape recorders, but will keep a complete log of all the edits made. It can also cue in other devices, such as caption generators and special-effects units, to carry out a set routine over a specified time.

Finally, time code techniques are the key to a great improvement in audio programming. Because multi-track audio tape recorders can be linked by the same time code system, it becomes possible to bring a much greater sophistication to the audio part of a video program.

Figure 18.4 A professional 1-inch edit suite. (Photo TVI Editing Ltd)

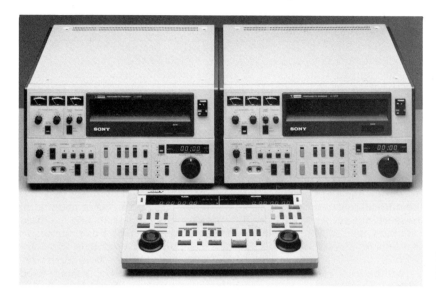

Figure 18.5 The workhorse of industrial video. A Sony Type V low band edit system.

The small studio

The above description of the potential sophistication of video production may seem designed to put off the small user altogether. So, in a way, it is. In general, anyone commissioning a video program should be happy that professional facilities can be used as required by the production. Nonetheless, there is a real demand for small in-house video facilities, especially in the training area.

If in-house use is to be confined to role-playing sequences and limited interest programs, it is probably best to:

☐ Base the system on low band U-Format if possible, especially if it may be necessary to have programs extended at outside facilities houses. Otherwise, industrial VHS should be chosen.

☐ Choose cameras on the basis of whether the majority use will be on location or in the studio.

☐ Limit the investment in control equipment to a simple edit controller and vision mixing system.

There are some large users of video who can justify the installation of a complete studio and production complex, but these users should either employ someone with the expertise to specify the system required, or engage the services of a consultant.

There is no reason for anyone running a training department to be afraid of using simple video equipment, provided that video is, in fact, the most appropriate medium for the task. Equally, it is not sensible to over-invest if there is no-one who will make full use of it. There are many horror stories of organizations with cupboards full of unused and out-of-date video equipment, a situation that probably arose out of the mistaken idea that video is all very easy, and that waving a camera in the direction of some activity will result in a usable program. Video is no different from the other AV media; it requires pre-planning, discipline, and a clear idea of its purpose.

Figure 18.6 The 'Maurice System' from CEL Electronics is one of the new generation of digital video effects units suitable for the commercial video studio.

Chapter 19

Video presentation systems

The television set is part of everyday life, and in recent years the widespread use of video recorders in the home has meant that the idea of audio-visual shows presented on a TV set is easily accepted, and apparently very simple.

There are, however, a few problems. The standard TV set is still a device that is best suited to the scale of the living room, so the AV user has to become aware of the compromises and alternative equipment that must be employed for larger audiences. The domestic user does not have to worry about TV signal standards; the commercial user does. This chapter reviews some of the points about television presentation that most affect the commercial and industrial video user.

TV signal standards

The problem of different signal standards is of most relevance to users who expect to receive video programs from abroad, or intend to send programs abroad. The evolution of color TV unfortunately resulted in the development of a number of different electrical signal formats being used to transmit the color picture information. To some extent, the different standards reflect improvements that were possible in second-generation equipment. There are also purely electrical reasons why there should be a difference between Europe and the USA. Unfortunately, additional standards have been introduced for political and financial reasons.

The first system into the field was the NTSC system, introduced in the USA in 1954. The letters stand for National Television Standards Committee or, unofficially, 'never twice the same color'. The system is used in the USA, Canada and Japan.

SECAM (Séquentiel Couleur à Mémoire), also known as the 'system essentially contrary to the American method', comes from France. It is used in France, USSR and most of eastern Europe.

Most of the rest of the world uses the German-developed PAL system, officially Phase Alternating Line, but also known as 'pictures at last'. Even within the main standards there are variations. SECAM has two versions, but fortunately only horizontal SECAM is now likely to be encountered. NTSC can be NTSC 3.58 or NTSC 4.43. The figures refer to the carrier frequency that is used to carry the color information.

NTSC 3.58 is the official version. The 4.43 modified version is derived in multi-standard videocassette recorders to allow the playback of standard NTSC tapes on PAL monitors. This highlights the awkward fact that usually the whole video chain must be of a particular standard. For multi-standard operation it is not usually sufficient to have just a multi-standard video recorder; the TV set must be multi-standard too.

The AV user fortunately only needs to be aware of the basic video signal standards, because the signal used should always be the simple video signal alone. When a TV signal is transmitted in a broadcast system, the variations between countries are considerable. However, this is a problem for the domestic TV set manufacturer rather than the AV user.

Videocassette standards

Most commercial users of video will encounter videotape in videocassettes. Unlike the audio business, where there is only one standard cassette, the video business has four widely used cassettes. The bad news is that there are new types, using narrower tape, coming soon. Three of these were essentially developed for the domestic market, and all use half-inch tape. The fourth uses three-quarter-inch tape.

VCR (Video Cassette Recorder) is a system developed by Philips. The cassette is also now referred to as V2000, the last version in the Philips system. It had a significant share of the market in continental Europe but is now obsolete.

VHS (Video Home System), developed by JVC, has the biggest share of the worldwide domestic market. Recently, a compact version of the VHS cassette has been introduced for portable equipment, called the C-Format. Betamax is the system developed by Sony, and is second to VHS in world popularity.

U-matic is the odd one out. This three-quarter-inch cassette system is not for domestic use. It was originally developed by Sony as a low-cost industrial video format. Over the years the performance has improved to such an extent that U-matic is used as the standard production medium for industrial AV. A variant called high band U-matic or BVU^{TM} reaches professional broadcast standards.

Out of all this confusion, what is relevant to the normal industrial user? Until recently the answer was easy. All industrial AV work would be done on U-matic, using the standard of the country concerned, e.g. PAL in the UK and NTSC in the USA. This is still good advice for the user looking for:

☐ The highest quality.

☐ The possible need to go to second- or even third-generation copies.

☐ Consistent replay quality on different machines.

☐ The international use of video programs.

The last item is particularly important. Most purchasers of U-matic machines would now buy multi-standard machines able to play PAL, SECAM and NTSC, thus facilitating the international exchange of programs. However, users should remember that the viewing monitor or video projector must also be able to deal with the different standards.

The great improvement in quality now available on the domestic formats has meant that they are taking an increasing share of the industrial market, on grounds of both cost and equipment size. For a program *made* to the highest possible quality, good results can be obtained on the domestic formats provided that each show copy cassette is made as a direct copy of the master. VHS is the leading half-inch format in the industrial market.

Conversion of color video signals from one standard to another requires very expensive equipment, and it is best to try to avoid the need for it. However, there are studios that provide standards conversion as a service, so tapes received from overseas that will not play on a user's equipment *can* be converted.

Figure 19.1 The decreasing size of videocassettes. In size order: U-matic, VHS and 8 mm.

There are two types of conversion service offered. One is to broadcast standard. This should be used whenever a program is intended to have wide distribution. For example, if a company in Europe makes a program in PAL, but wants to distribute 10, 50, 100 or 1000 copies in the USA, the best procedure is to take a high band or 1-inch copy master to the USA and have it converted to NTSC, which is also on high band. The resulting copy can then be used as the master for generating the bulk copies locally, which may well be on standard U-matic or VHS.

The second service is based on a low-definition standards converter, which has the advantage of convenience and low cost. However, its use should be limited to 'one to one' copying and dealing with emergencies.

8 mm systems

Whereas there is a degree of anarchy in half-inch tape formats (although, for many, VHS has become *the* half-inch showing standard), the major manufacturers *have* agreed on a universal 8 mm standard. Many users consider, justifiably, that they need another standard like a hole in the head. This accounts for its slow appearance on the market.

The target market for 8 mm systems is the domestic consumer, for home movie applications. It seems likely that there will be a slow introduction of 8 mm, because the latest compact VHS camera/recorders are nearly as small as the 8 mm equivalents, and have the great advantage of producing material that can immediately be played on the home VHS player. However, in the same way that domestic VHS has found its way into industrial applications, as a replacement market develops there may well be a place for 8 mm as a compact showing medium.

Figure 19.2 8 mm video equipment is very compact. Shown here are the 8 mm cassette, the separate video recorder and the remarkable Handycam™ combined camera and recorder. (Photo Sony)

Showing the picture

TV signals used in industrial AV are usually composite video signals that contain all the color picture information, but not sound. They have a bandwidth of about 5 MHz. TV pictures received by a domestic TV set are in the form of a modulated UHF signal, in which a color video and associated audio signal have been used to modulate a carrier frequency, typically 600 MHz.

Although it is possible to fit a video recorder with an RF modulator that will produce a UHF signal for direct connection to a domestic TV, this is not good practice because quality inevitably suffers. Thus, most industrial video programs are seen on monitors. 'Monitor' is a word that was once strictly reserved for the very expensive sets used by broadcasters to monitor the original picture, but it is now used to describe any set that has a video as opposed to RF signal input. Receiver monitors have both.

When a color TV picture is originated it starts life as three separate signals, one each for red, blue and green. When combined these give the whole color picture. Normally they are mixed together as a composite signal. However, there are some applications where it is best to keep them apart.

An RGB monitor is one which can deal with the separate signals. They are used particularly for the display of computer data.

The three signals must be synchronized together. In the absence of any other instruction, the green signal is used to synchronize the picture. Some systems, particularly those with multiple signal sources, use a separate sync signal.

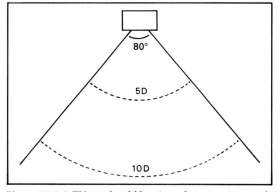

Figure 19.3 A TV set should be viewed at a maximum of 40 degrees off-axis, and at a distance of between five and ten times the screen diagonal.

Figure 19.4 A typical 66 cm (26-inch) presentation monitor. The clean lines make it easy to build into a system. (Photo Cameron Communications Ltd)

When choosing a monitor, the user must decide on:

□ *Size:* usually as big as possible for group viewing. The 26-inch monitor is the presentation room standard. Smaller monitors are used for individual viewing and portable equipment. The number of viewers who can easily watch monitors of various sizes is shown in Table 19.1.

□ *Loudspeaker:* in most fixed installations the audio will be played through a separate high-quality audio system, but sometimes the loudspeaker must be in the monitor.

□ *Standards:* is single-standard sufficient, or is there a need for multi-standard?

□ *RGB:* will the monitor be used for displaying data?

Table 19.1 Number of viewers of a single TV monitor (for general program material)

Screen	Diagonal	Number of viewers
9 in	9 in (230 mm)	4
12 in	12 in (305 mm)	7
15 in	15 in (380 mm)	10
19 in	19 in (483 mm)	16
21 in	21 in (533 mm)	20
23 in	23 in (584 mm)	24
25 in	25 in (635 mm)	28
26 in	26 in (660 mm)	31

The figures allow 0.75 sq m (8 sq ft) per viewer. Viewing distance is between five and ten times the screen diagonal. For viewing computer graphics and text material, the maximum viewing distance is about six screen diagonals (depending on the material) and viewing numbers are approximately halved.

□ *Receiver:* is a conventional TV receiver for off-air broadcasts required? Sometimes it is convenient to use a receiver monitor. In a multi-use system it can be better practice to use a separate TV tuner, especially when more than one monitor is in use.

□ Resolution: some applications, especially those calling for close viewing in laboratory and data display, require a high-resolution display.

Standard TV monitors with cathode-ray tube display are the most cost-effective way of showing TV pictures to small groups and individuals. They are less satisfactory for large groups. In some cases, the best solution to the problem is to use multiple monitors, sited so that everyone in the audience is within a satisfactory viewing distance of a monitor. This method is not recommended for serious viewing because the audience are all looking in different directions, and the apparent source of sound may not be the picture being looked at. The only answer is a big picture. The problem is that big video pictures are difficult to produce.

Video projection

Video projection is usually achieved by using three separate high-brightness cathode-ray tubes, one each for the red, green and blue. These tubes are about 12 cm (5 in) in diameter. Very large lenses must be used to collect the light and focus it on the screen. Electronic adjustments are made to the scanning signal to get the three colors to line up and to compensate for off-axis projection.

The most compact video projection systems use back projection to achieve an image of about 1.15 m (4 ft) diagonal. They use special screens to reject front-incident light, and are convenient to use because they are no more complicated than a large TV set.

Brighter images can be achieved by front projection. The lower-cost systems use a curved high-gain screen, and can only be viewed within a fairly narrow viewing angle. The screen and projector are one integrated optical system, so the resultant package is rather bulky. Pictures are typically 1.5 m (5 ft) diagonal.

The larger presentation room needs a system that can use a conventional screen, and can give a specified image size. Projectors for this application are available. In the dark these can give a picture up to 3 m (10 ft) wide. They can use conventional front or back projection screens, but benefit from a high-gain screen because their light output is less than half that of the conventional slide projector.

This type of video projector is only available with a one-lens system. This means that there is a

Figure 19.5 Back-projection TV sets are easier to install than their front-projection counterparts, but the ambient light conditions need to be right. (Photo Mitsubishi Electric (UK) Ltd)

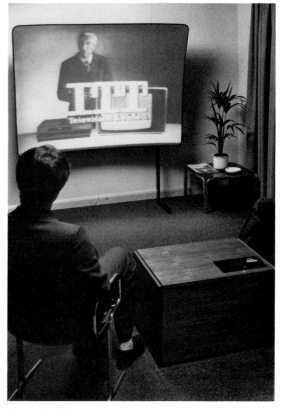

Figure 19.6 Front-projection television using a high-gain curved screen. (Photo ITT Consumer Products Ltd)

fixed distance from the screen for a given image size. Fortunately the projectors include electronic 'keystone elimination', so the projector can be installed on the ceiling or below the level of the bottom of the screen to allow unobstructed viewing. The projector can be made to project an electronically generated crosshatch pattern in each of the three colors. The usual procedure is to project the green one first, and to align it so that there is no distortion. The other two colors can then be lined up in turn to match the green.

The image size available from a video projector is limited by the amount of light generated by the cathode-ray tube. Attempts to increase this by using more beam current result in low definition and a short tube life. Because the standard video projector is by far the cheapest option, there have been several attempts to increase the light output in other ways. One manufacturer offers a four-tube projector with two green tubes to achieve a

Figure 19.7 Video projector with medium light output suitable for flat screens. This projector is suitable for multi-purpose presentation room use. (Photo Cameron Communications Ltd)

Figure 19.8 The presentation room at the Bristol, England, Division of British Aerospace Dynamics Group uses a ceiling-mounted video projector of the type shown in Figure 19.7.

950 lumen output, compared with a typical 440 lumens for a standard three-tube projector. A simpler method, costing no more, is to use two of the three-tube projectors directed at the same screen. This has the added advantage of providing a half-power emergency back-up if one projector fails, and is often used at important presentations.

Many years ago it was realized that the best answer to the problem of picture brightness would be to harness a conventional light source, such as a xenon arc lamp. This led to the development of special light valves, where a TV image is constructed by scanning a thin oil layer on a transparent plate. This varies the density of the

oil, and produces an image that can be projected like a movie film. The whole process must take place in a vacuum, and the oil film has to be continually reconstructed. A very expensive piece of equipment is required to do it.

The best example of the method in use is the Gretag Eidophor system. It is a flexible system, because, in principle, interchangeable lenses can be used. The largest machines have a light output of 7000 lumens, eighteen times as much as the best of the conventional video projectors, and can give pictures up to 13 m (43 ft) wide. An intermediate machine, both in terms of price and performance, is manufactured by General Electric of the USA. It features a single-gun system that allows for precise color registration and is easily set up. The General Electric Talaria™ projector is available in various models, with light outputs in the range of 500 to 2000 lumens.

Figure 19.10 The General Electric Talaria projector.

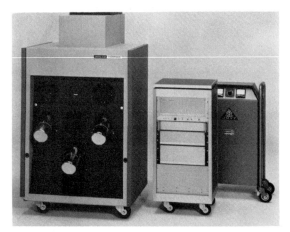

Figure 19.9 The Eidophor big-screen video projector. (Photo Gretag AG)

Another light-valve system originally developed for use in military briefing rooms, uses a liquid crystal light valve, which, although difficult to make, gives precision results. In the projectors developed by Hughes Aircraft, the light output of a xenon arc lamp is modulated by the output of a low-power cathode-ray tube via the liquid crystal light valve. At first these projectors were only used for graphics and text presentation, but they are now available to give full color video images, with a light output of 1000 lumens. If a low-cost projector using this technique could be manufactured, it would be an ideal presentation system component.

Figure 19.11 A video projector using the principle of laser scanning. It gives excellent 'black' performance, an infinite depth of focus, and is suitable for high-definition television. The test card being projected in this picture is 6 m (20 ft) wide. (Photo Dwight Cavendish Laboratories Ltd)

One other method, the use of a laser-scanning system to present the picture, may emerge as a serious contender. In principle, colored laser beams are directed by electronically controlled mirrors which cause the beams to scan the picture in a similar way to the electron beam in a cathode-ray tube. The system has the theoretical attraction of being independent of lens constraints, because the scanning arrangement can be programmed to produce the required image size over a wide range of projection distances.

The lasers needed to produce a bright picture are expensive, fragile and require water cooling. Even assuming that the present experimental systems are refined to a commercial product, laser-based video projectors are likely to find their main application in specialized permanent installations.

Flat-screen television

Ever since television first invaded the home there has been the dream of the flat-screen television that could be hung on the wall like a picture. This type of display was confidently predicted to be only a few years away in the 1950s, when the idea of a computer on a chip would have been laughed out of court. Now there is the microprocessor, but until very recently no real sign of the flat TV.

Flat single-color displays using a gas plasma are now appearing as viable display units for computers. It is possible that similar techniques could be extended to give a large screen with color. In EXPO 85, Matsushita demonstrated large color flat-screen displays using liquid crystal elements in front of ordinary fluorescent lamps. The 4 m × 3 m (13 ft × 10 ft) screens were made up from hundreds of 'bricks', each of which contained 256 three-color liquid crystal elements with their drive electronics. Although the brick structure of the screen was obtrusive, the demonstration did convincingly show that the technique could be developed for display purposes, if not for the home.

In 1985, electronics industry experts in Japan were predicting that the flat screen would become reality by 1990.

Recently, very large TV pictures of relatively low definition have been introduced to sports stadia and similar venues. Because these pictures must be viewed in daylight, and may be up to 15 m (50 ft) wide, a completely different technique must be used. These multi-million dollar displays use a matrix of colored fluorescent lamps or special display tubes to generate the picture. The best-known system is Mitsubishi's DiamondVision, but this has recently been joined by English Electric's StarVision.

Figure 19.12 The English Electric Starvision big-screen television display is used at major sporting events.

Determined not to be outdone, Sony at EXPO 85 in Japan showed a giant video screen with a 45 m (150 ft) screen, aptly called the Jumbotron. The sight of a 20 m (66 ft) high Sumo wrestler looming over the exhibition grounds was enough to terrify all but the most stout-hearted!

Although screens the size of the Jumbotron can only be permanently installed, smaller screens are available for rent for special occasions.

High-definition television

Much research is going on world-wide into high-definition television (HDTV). Present television standards build up a picture by scanning an electron beam across a cathode-ray tube face: 625 lines in PAL, and 525 lines in NTSC. In fact, the situation is a little more complicated than this. Each frame consists of two fields: 25 frames, 50 fields a second on PAL; 30 frames, 60 fields a second on NTSC. The fields are interlaced so, for example, on PAL a 625-line picture is built up by scanning the whole picture with half the available lines, and then filling in the gaps with the other half. For technical reasons, not all the available lines are used; field one carries picture information on lines 23 to 310, and field two uses lines 336 to 622.

In fact, the picture available on standard video is of an extraordinarily high quality which can only really be appreciated when seen on a broadcast standard monitor. Thus, in terms of producing a picture on a TV set up to, say, 66 cm (26 inch) tube size, the quality is limited not by the fundamental standard, but by the equipment used. For example, low-cost TV sets are not able to resolve more than about 300 lines, so there is little point in feeding them with a signal containing a lot more information.

The line structure of the video image starts to show up in video projection. It would also be noticeable if substantially larger tubes than the present sizes were to be introduced. Research on high-definition television has not confined itself to simply increasing the amount of picture information. It has also addressed the problem of picture format. The present TV aspect ratio of 4:3 was introduced both because it fitted on to the early round TV picture tubes and because it matched the film industry's Academy ratio. It is now accepted that a much more pleasing ratio would have a wide-screen look, like that already used in movie theaters.

In Japan, NHK, the Japanese broadcasting organization, has run an experimental high-definition television service in cooperation with the major equipment manufacturers. The system

Figure 19.13 A high-definition television set as used in the experimental NHK service in Japan. Notice the wide-screen format.

is based on an aspect ratio of 5:3 and uses 1125 lines. The picture tubes are enormous by present standards (see Figure 19.13) but the results are certainly spectacular. The high-definition video projection systems show up the format's potential to still greater advantage.

Despite the horrendous cost, it is likely that high-definition systems will find a number of specialized closed circuit applications, especially in fields such as medicine, prestige presentations, and even public entertainment. In a *small* movie theater the picture quality and brightness should eventually be able to match that from film.

It is equally unlikely that HDTV will, in this form, replace existing systems. This is because it requires a huge signal bandwidth, about 30 MHz as opposed to 5.5 MHz, and because it is totally incompatible with present equipment, which has such a large installed base. Researchers in Europe are paying much more attention to enhancing the present systems, arguing that, for most purposes, a bandwidth of about 6 MHz is theoretically capable of giving both higher quality and a wider picture. Their ideal is a system that is compatible with present receivers, but which carries extra picture information for those who have sets able to make use of it.

For the industrial AV user looking for high-definition pictures it is a matter of 'watch this space'!

Figure 19.14 A portable video unit combining videotape recorder and small monitor for one-to-one presentations. (Photo Tomorrow's World Developments Ltd)

The need often arises in one-to-one presentations, as discussed in Chapter 8. It is met by compact portable units that combine a videotape player with a small monitor. Although convenient for their intended application, they should *never* be used for larger groups.

The idea of the compact video projector with built-in videotape player will probably be developed, but even here users will have to be careful to ensure that the presentation environment and screen are suitable.

Video presentation units

Just as audio-visual presentation units are needed for slide presentations, there are some applications that need a video presentation unit.

Video systems

The needs of many video users are satisfied by a simple chain of one source and one display, e.g. a videotape player and monitor. As soon as multiple

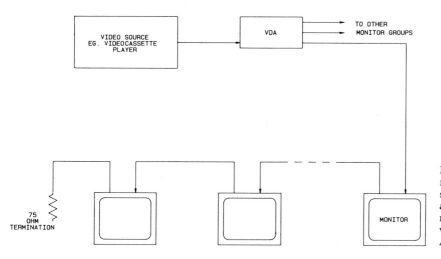

Figure 19.15 When several monitors are to be fed by the same signal they should have a loop-through facility, and may need feeding from a video distribution amplifier. Audio not shown.

Figure 19.16 A presentation room with many video sources will need remote-controlled video switching equipment.

displays, multiple sources, or both, are needed the user moves into the realms of video systems and will usually need a specialist video contractor to help specify and install the system. Some of the jargon that may be encountered now follows.

If a system uses several monitors, all driven from the same source, a video distribution amplifier or VDA will probably be needed. Most video sources can only deal with one or two monitors and the signal must be boosted for multiple monitor use. Some monitors have a loop-through facility that simplifies multiple-monitor installations (see Figure 19.15).

In a multi-purpose presentation room it will certainly be necessary for the video monitor or projector to be fed from several different sources; for example, VHS tape, U-matic tape, off air, and data display from one or more computers. To do this a video switcher is needed to select the source, and often it is convenient if it can be controlled remotely. Switchers are an essential component of the presentation room, but they normally give a rather dirty changeover from one source to another. The achievement of a clean cut without any picture roll is a much more complicated task. It requires all video sources to be synchronized and the use of a switcher that only switches in the vertical interval of the picture scan. These techniques are essential in video production, but only find their way into the highest-quality presentation installations.

Commonly it is necessary to display the output of a computer. Although personal computers used in the home often have a composite or RGB video

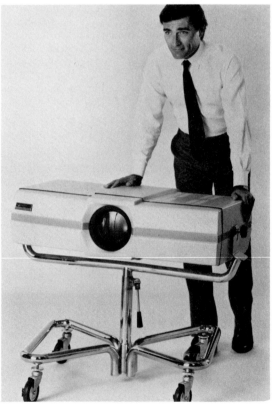

Figure 19.17 This video projector is suitable for video and computer data. Although it is a three-tube unit, a system of dichroic mirrors allows it to use only one lens. (Photo Electrohome Ltd (Canada))

output that is compatible with a standard monitor, more advanced computers use higher scanning rates to achieve a higher-definition image. Thus, it may be necessary to use a data projector to present computer displays to a large audience. Video projectors that can show standard video and color data, with a remote-controlled switch in scanning speeds, are available for most applications. However, there are some cases that can only be satisfactorily dealt with by dedicated monochrome or color data projectors.

Video conferencing

Sometimes two or more groups of people in separate locations use audio and video links to conduct a conference. The process requires specially equipped rooms. The participants must be arranged so that they can be seen by the cameras, and a special sound system must be installed to ensure clear transmission of sound without acoustic feedback or howlround. Arrangements must be made so that visual aid material can be directly fed into the system so it can be viewed by all the participants.

Not only is the installation expensive, but the cost of renting the broad-band links between the locations is very high. Slow-scan video systems are available: instead of sending a picture 25 or 30 times a second, they send one every few seconds. These systems can send pictures over ordinary telephone lines. Although superficially attractive

Figure 19.19 A demonstration of the video conferencing technique being used for distance teaching in the NEC Pavilion at EXPO 85. Here large projected images are being used.

for low-cost video conferencing, in practice they completely miss the point because conferences can only succeed if they work in real time. The enormous increase in international digital links is bringing the price per hour of link time down to a more affordable level, and within a few years the cost should no longer be a limitation.

Users of video conferencing need more than money to make a success of it. They must have a commitment to making the technique work for all the participants. This can really only happen when the participants are regular users of the system. Therefore, its main application is, and will remain, the support of regular management meetings. In theory a multinational company can hold a regular simultaneous board meeting in London, Melbourne and New York. In practice, time differences interfere with the longer-distance hook-ups.

Another use of video conferencing is distance teaching, where one lecturer is able to speak to two or more audiences at once. The extra dimension that the two-way link brings is that the remote audience can have a dialogue with the speaker, and he can see his questioners. This method of teaching is likely to be economic and effective in high-level training, where there are only a few expert teachers available, and in adult retraining.

Teletext

Teletext is a generic word for text-based information systems available on a public or network basis. Anyone who installs a video presentation system may require one of the teletext variants.

Figure 19.18 Video conferencing. British Telecom Videostream uses a system of data compression that allows full-color moving pictures of the participants and high-resolution graphics to be transmitted over digital links operating at between 768 Kbit/s and 2.048 Mbit/s.

The best known is broadcast teletext. This is a relatively limited information library containing up-to-the-minute public information, including news, weather reports, current events, travel information and stock market prices. The information reaches the viewer via a hidden part of the normal broadcast TV program, and is even used to produce subtitles for the program.

In the UK, the British Broadcasting Corporation broadcasts Ceefax and the Independent Broadcasting Authority broadcasts Oracle. Both are compatible, and only need a teletext adaptor to be added to a TV set, which is best bought as part of the original set. Selection of information pages is usually by an infra-red remote controller. Similar systems are available in other countries.

In a more powerful system the TV set is connected to the telephone network, giving access to a huge volume of information on central computers. Several 'viewdata' systems are in operation: in the UK, Prestel; in France, Antiope; in Canada Telidon; in the USA NAPLPS, among others.

With these viewdata systems, the user can be a passive viewer of information selected in the same way as on the broadcast systems, except that:

☐ The cost of the local telephone call must be paid.

☐ An extra fee may have to be paid to the information provider.

☐ The choice of information available is much greater. In principle, there is no technical limit.

Alternatively, and this is the big difference, the system allows interactive use. For example, the user can become the information provider, or can, in principle, use the system for reserving or purchasing tickets and goods.

In practice, the public viewdata network tends to the passive, although intense marketing efforts are being made to change this. On the other hand, *private* viewdata networks, especially in the travel and financial trades, are already proving a great success. These are fully interactive.

The videodisk

The videodisk has a curious history. It was originally invented for the consumer market, where, it was thought, video LPs would be a more acceptable method of distributing entertainment programs than tape. The advent of the low-cost home video recorder and the realization that this could also be used for the timeshift recording of broadcast material made the videodisk player very much a second option. The videodisk has never

Figure 19.20 A Laservision™ videodisk.

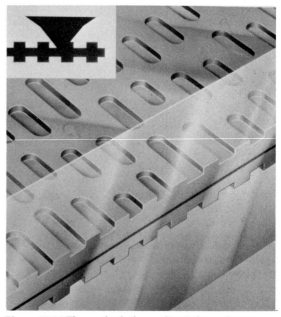

Figure 19.21 The method of recording information on a laserdisk. (Diagram by Philips)

really got off the ground as a domestic item, although eventually it may become more popular as a direct result of experience gained with its successful use in industry and training.

Many of the attributes of videodisks have already been discussed in Chapter 12. Videodisks bear some similarity to the long-playing phonograph record, but the 'grooves' are very much closer together, and the information is recorded as pits or variations in the depth of the 'groove'. In fact, the disks do not have actual grooves, except in the now-defunct CED system, and the pit pattern is laid down concentrically as shown in Figure 19.21 or, in some record/play units, spirally. The pick-up system includes sophisticated electronics to ensure that it is directly over the required part of the disk. The information is read in one of two ways.

In the capacitance disk, a stylus is in contact with the disk, and an electrode picks up the information by detecting the rapid changes of capacitance resulting from the variation in depth. In the laserdisk the surface is reflective, and the signal is derived by scanning the surface with a tiny light beam produced by a small laser. The variation in the reflected light is detected by a photocell which produces the corresponding electrical signals.

The laserdisk is fairly robust. Although it should still be handled with care, the actual playing surface is protected by an outer transparent layer; the laser is focused on the information layer beneath it (see Figure 19.22). The capacitance disk is less robust, and for this reason is normally kept in a protective jacket which can only be removed once the disk is in the player.

Figure 19.22 The scanning of a laserdisk. (Diagram by Philips)

The characteristics of available disks are shown in Table 19.2, the technology of recordable disks is fluid, and there may be some changes as the technology develops. The originators of capacitance disks were RCA with their Selectavision™. They have now withdrawn from the market, leaving the field to JVC with their Video High Density or VHD system, so Table 19.2 only refers to VHD.

The laserdisk was invented by Philips, which is the major source of players for the PAL market. Major Japanese manufacturers include Pioneer, Sony and Hitachi. The laserdisk is either CAV or CLV (constant angular velocity or constant linear velocity). In CAV players the disk rotates at a constant speed, and one revolution of the disk corresponds exactly to one frame (two fields). In CLV the rotation speed varies, but the speed of the pick-up over the disk surface is constant. This allows more information to be packed in on the

Table 19.2 Characteristics of videodisk systems

	Laser CAV	Laser CLV	Optical disk recorder (ODR)	Video high density (VHD)
Disk size	30 cm (12 in)	30 cm (12 in)	20 cm (8 in)	25 cm (10 in)
Frames per side PAL	54 000	87 000		90 000
Frames per side NTSC	54 000	108 000	24 000	108 000
Frames per revolution	1	variable		2
Disk speed, PAL	1500 rev/min	variable		750 rev/min
Disk speed, NTSC	1800 rev/min	variable		900 rev/min
Playing time per side, PAL	36 min	58 min		60 min
Playing time per side, NTSC	30 min	60 min	13.3 min	60 min
Audio tracks	2	2	2	2

Notes: 8 inch laserdisks are available. The still-frame capacity of VHD is half the number of frames shown (which is the moving image capacity).

outer parts of the disk, and hence a longer playing time.

The result of the one frame per revolution in the CAV disks is that it is very easy to select an individual frame. CLV cannot really be used for still-frame work. The video high density system (VHD) can also be used for stills because it is a constant-speed system. However, it shows two frames per revolution, so still frames must be recorded as stills with at least two identical images. Thus, the still-frame capacity of VHD is 45,000 images in PAL.

Videodisks are now essentially a replay medium. The producer must make a master tape which is used by the disk manufacturer to make a master disk stamper, which is in turn used to make as many disks as required. Disk mastering charges may seem high, but with careful planning they are reasonable as a proportion of any particular project, especially because the disk opens up applications of video that cannot be met any other way. The cost of extra disks once the master has been made is very reasonable. The disk manufacturers are working hard to give a fast turn-round for users, including most industrial users, who need a few disks quickly. Turn-rounds of a few days are quite normal.

The needs of many one-of-a-kind installations, and of those who develop program material, would be better met by recordable videodisks. These are now available, but at a high price. In Table 19.2 the data for the optical disk recorder (ODR) is for a unit made by Panasonic. Teac make a recorder whose disk has the same capacity as the standard laserdisk. This type of equipment will probably be available at relatively low prices in the near future.

At present the industrial market has a good choice of playback machines. These vary in sophistication depending on the degree of control required, as discussed in Chapter 12. The general rule is, as always, to use the simplest equipment that meets the needs of the job. Industrial users should definitely *not* use disk players intended for the domestic market. These are not designed for continuous use.

Multi-screen video

A display medium that has recently gained some popularity is multi-screen video, where a number of monitors operate in synchronization. The best-known example of the technique is the video wall where banks of monitors are arranged in a regular pattern, usually 3 × 3, 6 × 6 or some

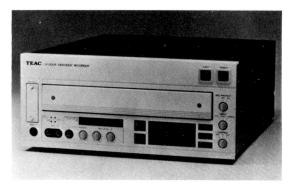

Figure 19.23 The recordable laserdisk will open up many new AV applications. The Teac LV-200A is one of the new recorders able to record 54,000 video frames, individually or in moving image sequences.

intermediate array, but sometimes in groups of hundreds of monitors. However, the technique can also be used in more subtle ways, which makes it especially suitable for supporting multi-media exhibits.

There are many different ways of achieving multi-screen video; the method actually chosen will depend on the desired effect, available production resources, and budget. The problems are best understood by reference to an example. Often a nine-monitor array arranged as a 3×3 matrix is required. A matrix with the same number of monitors vertically as horizontally has the advantage that the overall screen ratio is the same as that for a single screen.

It is likely that the user will want one image to spread across all nine monitors. There are two ways of achieving this. The array can be fed from a single picture source, and a black box forming part of the show system is used to split the picture on site. Otherwise, the array can be fed from nine video players (either disk or tape) which run in frame sync; each of the players carries the part of the image corresponding to one monitor.

Both methods are expensive! The multiple-player approach is best for long-running or permanent exhibits with relatively short program lengths. It allows total flexibility in the way that the show is made up, ranging from a different moving picture on every screen, to one picture split across all screens. All production is done in the studio, so the quality can be of the highest standard. In particular, splits across many screens can be done on broadcast-standard equipment and so can be made free of obvious pixellation.

In fact, if splitting is to be done across more than nine screens it is recommended that material is

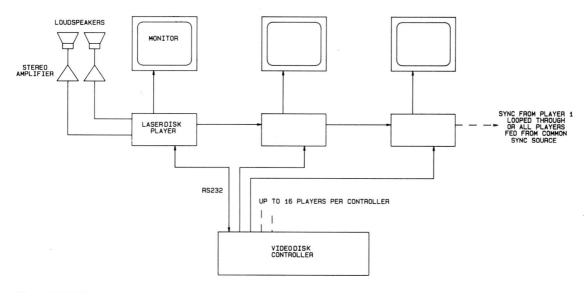

Figure 19.24 The principle of multi-screen video using multiple videodisk players.

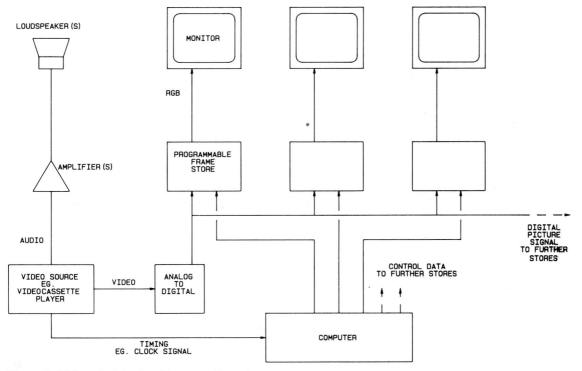

Figure 19.25 The principle of multi-screen video using multiple frame stores.

originated on 35 mm movie film, and first optically split. For example, a 6 × 6 array would have its original film optically split into four 35 mm films. These would be transferred to 1-inch videotape, and each, in turn, split nine ways using a broadcast-standard special-effects unit.

It is possible to run the final show from videotape players. However, these must have capstan servo control and some system for precision start location. This means editing machines and time code equipment, possibly justified for a special occasion, but uneconomic and requiring too much maintenance to be used for a permanent show.

Multiple videodisk players represent the answer for this method. The system needs top-of-the-range players that have provision for external sync and computer control, but the complete system architecture is quite simple as shown in Figure 19.24. The central controller allows for both one-shot and continuous showing. In continuous shows the interval between shows is only two or three seconds, and in the one-shot use the system can hold on a still frame when it is not actually running a show.

Although the multiple-disk system gives by far the best results it may not be practical for the user who wants:

☐ long shows;

☐ frequent program changes;

☐ the ability to feed in existing program material;

☐ the ability to feed in live images.

Figure 19.26 A neat disk-based multi-screen video display at EXPO 85.

Figure 19.27 This 100-monitor display in the Korean Pavilion at EXPO 85 used a combination of frame stores and programmed image switching. The maximum split was over sixteen monitors (4 × 4), although this is not a theoretical limitation. Systems allowing a split across 256 monitors (16 × 16) are available.

In these cases the 'black box' solution must be used. This works by digitizing the picture and splitting it mathematically. Each monitor must have a digital frame store to process its part of the overall image. Depending on facilities and image quality these systems have a price which can be higher or lower than that of the multiple disk system. In general, the cost of originating programs will be less, especially if they are based on existing material.

The simplest systems split an incoming signal across the requisite number of screens. However, this is a bit dull, so more sophisticated units allow the on-site control computer to be programmed to manipulate the display in various ways. A typical array might have two moving-picture inputs, and the computer allows each screen to show:

☐ either of the two original pictures;

☐ a freeze frame;

☐ part of a split image of one of the original pictures;

☐ black or color wash.

In this way, some variety can be introduced. Although it is not possible to have nine different moving pictures on the nine-screen array, it is possible to have nine different still pictures by successive freeze framing. There obviously cannot

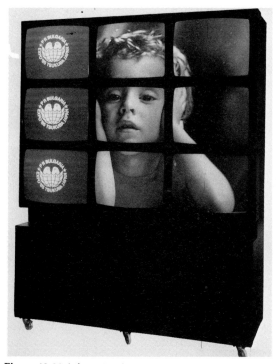

be more moving pictures than there are sources –
in this example, two. The whole show production
is divided into two parts: the origination of the
prime material, and the separate programming of
the playback system.

In between the multiple-player system and the
programmed frame-store system there are several
other possibilities, including the use of programmed switchers. Often, for the sake of economy,
monitor arrays that repeat the main image pattern
two or three times are built.

Multi-screen video is *not* synonymous with the
video wall. For some users, a few screens served
by video projectors would represent a better use of
resources, especially because there would be no
gaps between the screens. Although efforts are
being made to eliminate the gaps on multiple
monitors, by using fresnel lenses or fiber optics,
these systems are not entirely satisfactory because
they result in a directional image.

Another possibility is the use of separated, but
synchronized, monitors set within an exhibit. If
the multiple-disk system is used, the disk control
computer can also carry other programming
information to operate lighting, slide projection
and animation in synchronization with the video
program.

Comments on multi-screen video have so far

Figure 19.28 A free-standing nine-monitor display.

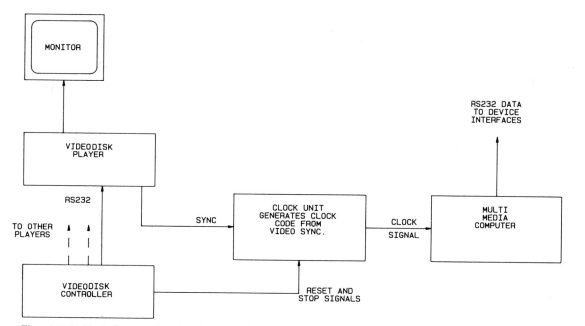

Figure 19.29 Block diagram showing the principle of
linking videodisks to multi-media displays.

Figure 19.30 This exhibit in the Toshiba Pavilion at EXPO 85 linked the activities of seven videodisk players to the programmed operation of a banknote-sorting machine and associated lighting.

assumed the use of a linear program, but this is not a necessary feature. Disk-fed systems lend themselves to interactivity, where either the whole display changes program direction in response to an external stimulus, or parts of it show different information according to audience input. A large-scale example of the principle was the NEC Computers and Communications Theater at EXPO 85. Here a 27-screen show using video projection with 2.5 m (8 ft) screens was based on multiple videodisk players. Each member of the audience had a touch-screen response system that allowed them to choose the way they wished the presentation to go (in this case, an imaginary space journey). The use of computer-controlled synchronized disks allowed instant changes in program direction.

Magnetic disks

A recent innovation, first promoted by Sony, is the ability to record individual images onto magnetic disks, similar to the small disks used in personal computers. Here it is possible to transfer any single-frame video image, or, with a suitable recorder, any photographic image, to a magnetic disk holding, for example, 50 pictures.

The equipment, although not yet at a consumer price level, is easily affordable by the industrial AV user. It is clear that such equipment will simplify small audience presentations where video equipment is needed anyway. However, serious presenters will continue to use slides on their own where large audiences or highest still-image quality are required.

Chapter 20

Computers in audio-visual

Computers are playing an increasing part in the world of audio-visual, sometimes helping to do things better, sometimes as the key to completely new applications. This short chapter reviews their impact and draws attention to a few items that do not easily fit under the other chapter headings.

Computers are now taken for granted in audio-visual work, and are used in:

☐ Computer typesetting.

☐ Computerized rostrum camera control.

☐ Multi-image programming (see Chapter 16).

☐ Computer graphics and the production of slides by computer (see Chapter 3).

☐ Computer animation in video and film production.

☐ Computer control of video editing (see Chapter 18).

☐ Interactive AV programs (see Chapter 12).

☐ Display of data held in the computer.

Rostrum camera control

The rostrum camera, mentioned in Chapter 2, is the key to the production of high-quality slides. Rostrum cameras have a number of mechanical movements; their precise application produces either a special effect or a picture sequence. On a slide rostrum camera a streak effect can be produced by precise movement of the artwork or transparency which is being copied during a time exposure. This movement can be done by hand, but is much more controllable and repeatable if controlled by a computer. Other examples are 'step and repeat' exposures to make several slides from one original. Computer control can both increase the throughput of the rostrum camera, and introduce effects that are simply not practical with manual control.

Computer control is especially important for the recently introduced video rostrum cameras be-

cause it is as relevant (if not more) to the production of moving pictures, both movie and video, as it is to still photography. Computer control can greatly speed up the production of animation sequences, and create much smoother results when complex camera movements are needed. The high-speed combination of panning, rotating and zooming then becomes practical.

Computer animation

The direct creation of animated moving pictures by computer is a subject of intense study, and is used increasingly. However, the technique is only accessible and of relevance to the professional film maker, so is not of great concern to the AV user, except as part of a commissioned production.

Display of computer data

Chapter 3 reviewed the role of computer graphics, with special reference to the creation of business graphics, usually involving the production of charts and graphs. Here the data to be displayed were provided by some external source, probably a piece of paper with the relevant data.

Often the data will have been derived from computer-stored data. Indeed, the piece of paper may well be the result of someone copying some information from a computer screen, or be derived from a computer printout.

A normal computer display screen is quite unsuitable for presentation to a group. It contains far too much information and is illegible unless viewed from a short distance. Often a slide must combine the information held on several different pages of data held in the computer.

In the computer world it has long been recognized that the presentation of data in a

Figure 20.1 The Telex Magnabyte TM allows the monochrome display output of a personal computer to be shown to groups using an overhead projector as a light source. It uses a large liquid crystal array. (Photo Gordon Audio Visual Ltd)

Figure 20.2 Computer programs are available to reformat data held in mainframe computers into a form suitable for presentation. (Photo ISSCO)

digestible form requires a special computer program. Companies such as ISSCO™ and SAS Institute™ offer computer software designed to reside in customers' mainframe computers, which allow the creation of business graphics based on corporate data held in the same computer. Until recently the resulting display has usually been viewed on a color monitor, or on a hard copy printout. Now it is often necessary to use a film recorder to make a slide for group viewing.

Direct computer display

It is possible to create graphics images, and then, instead of making a slide, to present them in sequence directly from the computer (see Chapter 3).

Figure 20.3 The Videoshow 150™, a special-purpose 'black box' for the sequential display of computer prepared graphics. (Photo Mediatech Ltd)

It is important to realize that the direct presentation of computer graphics implies a quality limitation. This limitation will be quite acceptable, and will probably be unnoticeable, if the sequence is viewed on a video monitor. If the image is to be presented to a larger group, an expensive video projector will be needed, and the quality limitation may become an important factor.

Assuming that a pure video presentation is acceptable, the graphics computer must have the ability not only to create images, but also to store them so that they can easily be shuffled into the required order. It should be no more difficult to give a sequential image presentation than to use a slide projector.

Various possibilities exist. At one extreme there are complex systems used in professional video production that can access and display images in any required order, with the availability of special effects. More realistically, at the other extreme there are systems that permit the creation of simple graphics on a personal computer, which can then be transferred to a black box. This black box is like an electronic slide projector, and allows both sequential and random-access display of computer graphic images, but not photographic images.

These devices must have buffer memories. A single video image requires a considerable amount of memory, and in a low-cost system it can take an appreciable time to write a new page, especially if the data must be retrieved from a floppy disk. Thus, there is usually at least one additional frame store, which can be loaded with a new image while the previous image is being displayed.

Some users are now using these types of system as the basis for speaker support at sales conferences. For small audiences who need only a relatively small single screen the technique is valid. However, some disadvantages of its use include poor image quality, and much greater cost. A video projector and computer are 10–50 times as expensive as a slide projector. Advantages include the fact that images can be changed right up to the last minute, and the fact that certain types of animation, such as the ability to draw pictures and do more elaborate build-ups, are available. This is not easily done with slides.

Data projector

For those who must present the contents of a standard personal computer monochrome monitor screen to a small group, the liquid-crystal-based data projector is available. It comes in two forms: an overlay to a standard overhead projector (OHP), or a compact self-contained projector (Figure 20.4).

The image presented is bright, but of low contrast. Unless a graphics program is used to create large characters, the display requires a short viewing distance because the individual character sizes are small. Users should establish beforehand that the results obtained from the data projector are satisfactory for the intended audience arrangement.

Figure 20.4 The Kodak Datashow™ projector can project a high-brightness image. It is suitable for projecting suitably formatted computer data on to standard projection screens.

Chapter 21

Lenses

All audio-visual systems using projection finally rely on an objective lens to focus the image on the screen. The variety of lenses available is what makes projection so flexible. The same principles apply to the projection of movie, slides or video. However, it is in slide projection that the versatility of different lenses is best demonstrated. This chapter will first describe the possibilities for slide projection, and then show how movie and video projection differ.

Basic lens properties

It is best first to explain some technical terms relating to lenses, so that their importance can be understood when choosing a lens.

The *focal length* of a lens is the distance from the plane in which the lens forms an image of objects at infinity to the node of emission. This textbook description disguises the information that matters; long focal length projection lenses are needed for long-distance projection, and short focal length lenses for short-distance projection. The formula relating projection distance to image size is easy to remember. A 35 mm lens projecting a 35 mm slide will give a picture that is as wide as the projection distance, e.g. a 2 m (6.5 ft) picture at 2 m projection distance. Multiply the focal length by 2 and the projection distance doubles, i.e. a 70 mm lens gives a 2 m (6.5 ft) picture at 4 m (13 ft) throw; a 140 mm lens gives it at 8 m (26 ft), and so on. The practical range of projection lenses for 35 mm slide projection is 25 mm to 400 mm, although there is a very expensive 15 mm lens available, and special applications can use focal lengths up to 1000 mm. This is summarized in Table 21.1.

The *relative aperture* of a lens is an expression of its ability to collect, and therefore project, light. Technically, it is the ratio between the effective diameter of the lens and its focal length. The smaller this ratio, the more light passed. The same applies to lenses used for picture taking, and for this reason lenses with low relative apertures of *f*-numbers are called faster lenses. It is difficult to make long focal length lenses with low *f*-numbers because the diameter must increase and they get very big. In slide projection short focal length lenses should be *f*/2.8. With the optical system of standard slide projectors there is no real benefit in having longer focal length lenses with better than *f*/3.5, because the cone of light from the light source is not big enough to take advantage of faster lenses. With lenses of focal length greater than 300 mm it is difficult to get lenses as fast as *f*/3.5.

Modern high-quality projection lenses will give an image that is sharp to the corners, that is free of

---

Table 21.1 The standard lens formula

The full formula is:

$$W = \frac{D}{f - 1} \times w$$

where W is the screen width
  w is the film aperture width
  D is the projection distance measured from the image nodal point of the lens to the screen
  f is the focal length of the lens

However, for all practical purposes a much simpler formula can be used:

$$\frac{\text{Screen dimension}}{\text{Film frame dimension}} = \frac{\text{Screen distance}}{\text{Lens focal length}}$$

For slides, the film aperture width is 35 mm. Provided the focal length of the lens is also expressed in mm, the picture width of the projection distance can be easily calculated. It does not matter whether this is done in meters or feet, provided the same units are used for both.

---

chromatic aberration (color fringing) and of *spherical aberration* (the tendency to 'pin cushion'). They should also be *multicoated*. The lens includes at least three, and more usually five or six or as many as nine elements. At each glass/air surface there is some reflection, and therefore wasting, of light. For example, if ten per cent of the light was lost at each element of a nine-element lens, the lens would lose more than sixty per cent of the incident light. A sophisticated technique of vacuum coating the lens elements eliminates this problem. As much as twenty five per cent extra light is given this way. A coated lens can be distinguished by a bluish bloom on the lens. Care should be taken not to touch the lens surface and to clean it as carefully as an expensive camera lens.

Lenses for slide projection

Table 21.2 shows a list of strandard slide-projection lenses available from one manufacturer, and gives a good idea of the wide range available.

In general, a permanent installation with a fixed projection distance is best served by the use of the appropriate fixed focal length lens because these

---

Table 21.2 Slide-projection lenses

Fixed focal length, front projection

90 mm *f*/2.4	100 mm *f*/2.8 or *f*/3.5
150 mm *f*/3.5	120 mm *f*/2.8 or *f*/3.5
180 mm *f*/3.5	150 mm *f*/2.8
250 mm *f*/4.3	200 mm *f*/3.8
	250 mm *f*/3.8
	300 mm *f*/4.5
	350 mm *f*/5.2
	400 mm *f*/3.8

Zoom lenses

70–125 mm *f*/3.5
110–200 mm *f*/3.4
200–300 mm *f*/3.5

Rear projection lenses

25 mm *f*/2.8	15 mm *f*/2.8 (very special)
35 mm *f*/2.8	37 mm *f*/2.8
45 mm *f*/2.8	50 mm *f*/2.8
60 mm *f*/2.8	

Perspective control lenses

45 mm *f*/2.8	60 mm *f*/2.8 (Extra offset)
60 mm *f*/2.8	
90 mm *f*/2.8	
70–125 mm *f*3.5	

Table 21.3 Projected apertures

Image	Width (mm)	Height (mm)
Super 8 movie	5.46	4.01
16 mm movie	9.65	7.26
35 mm movie Academy	21.11	15.29
35 mm movie widescreen 1.85:1	21.11	11.41
35 mm movie anamorphic	21.29	18.29
70 mm movie	48.59	22.10
Superslide	37.3	37.3
35 mm slide ANSI standard	34.2	22.9
35 mm slide WESS No 2	34.8	23.4
35 mm filmstrip	22.3	16.7

---

are less expensive than zoom lenses and usually give more light. However, some of the modern zoom projection lenses are so good that they can be used in permanent installations when there is no fixed lens suitable. Although the lens manufacturers will make special lenses to order, it is not an economic proposition unless several hundred lenses are required.

Table 21.3 draws attention to the fact that there can be slight differences between slide mounts. This will affect the image size and must be taken into account when designing precision installations. The differences arise because the actual image on the film occupies a standard area which the mount deliberately masks slightly to ensure a perfectly projected picture. Early standards for the masking were somewhat pessimistic, so slide mount manufacturers increased the area to the maximum practicable.

Zoom lenses

Zoom lenses are lenses with an adjustable focal length. They are invaluable for travelling AV shows because they allow precise filling of the screen from the available projection distance. It is important that these lenses should not introduce any distortion, especially in multivision applications. For economic and practical reasons the zoom range is not usually greater than 2:1. However, as can be seen from Table 21.4, just three lenses can allow zooming all the way from 70 mm to 300 mm. Most AV presentation work using front projection can be based on the 110–200 mm lens.

Professional-quality zoom lenses are all supplied in metal barrels, and have a zoom lock that locks the zoom action. This is an essential feature for multi-image work.

Figure 21.1 This range of zoom lenses zoom from 70 mm to 300 mm. The 200–300 mm lens must have a large diameter to achieve its relative aperture of *f*/3.5.

It is technically possible, but also rather expensive, to have motorized zoom lenses that correctly maintain focus while zooming. The effect is of limited value except to specialist users like planetaria. Some of these lenses are based on camera lenses, and either have limited light output, or only cover half a slide frame, or both. Some special lens systems have been made with a 20:1 zoom ratio, but their light output is very low.

Back-projection lenses

Although the best picture on back projection is achieved by using a lens with a long focal length, normally users want to use as little space as possible for back projection. A useful range of lenses is available to meet applications varying from compact single-projector display cabinets to

Figure 21.2 Typical fixed-focal-length lenses. The 400 mm is the longest focal length used in normal auditorium applications, but longer 35 mm slide-projection lenses have been used in planetaria and other specialized installations.

Table 21.4 Examples of slide-projection distances using different lenses

Focal length	Picture width, m (ft)										
	1 m	(3.3 ft)	1.5 m	(5 ft)	2 m	(6.6 ft)	2.5 m	(8.2 ft)	3 m	(10 ft)	3.5 m	(11.5 ft)
25 mm	0.7	(2.3)	1.1	(3.6)	1.4	(4.6)	1.8	(5.9)	2.1	(6.9)	2.5	(8.2
35 mm	1.0	(3.3)	1.5	(5.0)	2.0	(6.6)	2.5	(8.2)	3.0	(9.8)	3.5	(11.5)
45 mm	1.3	(4.3)	1.9	(6.2)	2.6	(8.5)	3.2	(10.5)	3.9	(12.8)	4.5	(14.8)
60 mm	1.7	(5.6)	2.6	(8.5)	3.4	(11.2)	4.3	(14.1)	5.1	(16.7)	6.0	(19.7)
90 mm	2.6	(8.5)	3.9	(12.8)	5.1	(16.7)	6.4	(21.0)	7.7	(25.3)	9.0	(29.5)
110 mm	3.1	(10.2)	4.7	(19.4)	6.3	(20.7)	7.9	(25.9)	9.4	(30.8)	11.0	(36.1)
150 mm	4.3	(14.1)	6.9	(21.0)	8.6	(28.2)	10.7	(35.1)	12.9	(42.3)	15.0	(49.2)
180 mm	5.1	(16.7)	7.7	(25.3)	10.3	(33.8)	12.9	(42.3)	15.4	(50.5)	18.0	(59.1)
200 mm	5.7	(18.7)	8.6	(28.2)	11.4	(37.4)	14.3	(46.9)	17.1	(56.1)	20.0	(65.6)
250 mm	7.1	(23.3)	10.7	(35.1)	14.3	(46.9)	17.9	(58.7)	21.4	(70.2)	25.0	(82.0)
300 mm	8.6	(28.2)	12.6	(41.3)	17.1	(56.1)	21.4	(70.2)	25.7	(84.3)	30.0	(98.4)

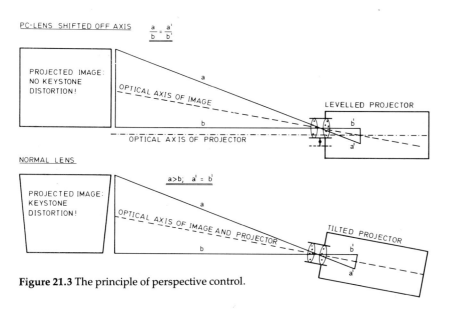

Figure 21.3 The principle of perspective control.

complex board room or presentation room systems using several projectors.

A problem arises when more than one projector is needed to serve the same screen area. This will always happen when the dissolve or multi-image technique is used. Inevitably, only one projector can be perfectly lined up with the screen; more usually, a compromise set-up is used so that none of the projectors is exactly lined up. This results in keystone distortion, where the image is not properly rectilinear. It shows up particularly badly in graphics sequences.

The solution to the problem is the perspective control lens, in which the optical axis of the lens is adjustable. This can effectively eliminate the problem for all pictures greater than 1.8 m (6 ft) wide, and considerably reduce it for smaller pictures. In fact there is a special 60 mm lens with an extra offset that eliminates the problem on pictures as small as 1 m (3.3 ft) wide.

Fixed-focal-length perspective control lenses are often supplied with their exact focal length marked. This means that in a multi-image presentation it is possible to select lenses that match exactly for each image area. Professional lenses are usually guaranteed to be within 0.5 per cent of their effective focal length.

Anamorphic lenses

An anamorphic lens is one which, when used as a camera lens, squeezes a wider than normal picture onto the film. When used as a projection lens it unsqueezes the picture.

The technique is usually used for movie film. The first major commercial exploitation was called Cinemascope™. Typically, the anamorphic system on movie projection is able to give a picture ratio of 2.4:1, as opposed to the normal widescreen ratio of about 1.8:1.

Figure 21.4 Perspective control lenses. These eliminate keystone distortion. They are available in focal lengths from 45 mm to 125 mm.

The same technique can be used on slides. In this case, the stretching is to about 2:1 from the normal 1.5:1. In both cases the anamorphic lens is in addition to the prime lens.

The anamorphic lens system for use with slides is Iscorama™ made by Isco Optik of West Germany. They can also provide professional copying lenses to allow anamorphic slide production on a rostrum camera.

Movie lenses

All the rules which apply to slide projection lenses also apply, where relevant, to movie projection. Notice that the film aperture is different, so that the focal length needed to produce a given picture size will be different than that which applies to slides.

For example, on 16 mm film the width of the actual film aperture is 9.65 mm. A 50 mm lens will give a 3 m (10 ft) wide picture at a projection distance of about 15.5 m (50 ft). Most 16 mm projection is done with fixed-focal-length lenses, although zooms are available. If movie film is back-projected it is necessary to use a reversing mirror; although it is possible to turn a slide around for back projection, this cannot be done with a movie film because of the position of the sound track. The reversing mirror can either be external to the lens, or can actually be built into the lens. Lenses down to 9.5 mm are available, to give a picture width the same as the projection distance.

Lenses for 35 mm movie projection are usually made with a small relative aperture, to ensure maximum possible light on the screen, and to match the professional illumination systems used.

Figure 21.5 A short focal length lens for rear projection of 16 mm movie. It uses a built-in reversing mirror. (Photo Buhl Optical)

Figure 21.6 An anamorphic lens as used in professional cinema projection. (Photo Isco Optic)

In principle, it is possible to put more light through a frame of movie film (compared to a slide) because an individual frame is only in the gate for a fraction of a second. Against this, the shutter system is working to reduce the light available.

35 mm movie projection lenses are available in a wide range of focal lengths, in as close as 5 mm increments. Fixed-focal-length lenses are always used, and the final matching of the image to a particular screen is done by slight adjustments to the aperture plate. This is a metal mask in the projection gate that would normally be the standard projection size, but can be obtained slightly undersize, and then enlarged to match the particular installation.

Lenses for 35 mm movie projection are designed with auditorium applications in mind. Lenses for special display applications which require short projection distances must be specially made, or be adapted from slide-projection lenses with a considerable loss in light output.

Projection of 70 mm movies is normally only done in premier cinemas, but it can also be required in special displays, especially visitor entertainment in theme parks. A full range of lenses is available for conventional applications, as well as a few high-cost lenses to produce special effects. Some examples include a 50 mm lens to give a very big image at short projection distance, and very short focal length lenses to fill a whole dome with a single picture. When these lenses are used, the picture must be very steady and not all projectors are suitable.

Video-projection lenses

Video projection is much less flexible than slide or movie projection. However, there are a few more flexible systems such as the Eidophor system which uses light valves. These systems do permit a choice of lenses, and are sufficiently expensive that the lens problem is relatively insignificant. Lenses can be found for most applications.

However, most commercial users of video projection use video projectors based on standard cathode-ray tubes. The normal arrangement is three tubes (red, green and blue). The problem is to get the light efficiently from the surface of these tubes to the screen.

So little light is available in these systems that it is really only feasible to use fixed-focal-length lenses, and to move the projector to the position that gives the required picture size. The lenses must be $f/1$ and must deal with an input aperture equivalent to a 12 cm (5 in) tube. This results in a short projection distance, typically 1.5 times the screen width. Fortunately, it is practicable to mount the projector off axis because keystone distortion can be corrected electronically.

Projection lenses in video projectors are usually made of plastic. However, projectors that are intended for projecting computer data and graphics greatly benefit from the use of glass lenses, and these should always be specified for such installations.

Chapter 22

Screens

Most AV methods described in this book involve the projected image. In the end the image must be viewed on a screen. This chapter reviews the different types of screen available, and also the reasons for choosing different projection methods.

The concept of screen gain

Some projection screens seem brighter than others, and although it might seem best to use the brightest possible screen for all applications, there could be some disadvantages. To understand why, it is necessary to understand what is meant by 'screen gain' and how it is achieved.

The unit of 'luminous flux', or light output from the projector, is the *lumen*. The 'illumination' of a surface can be described as the luminous flux per unit area; it is measured in lumens per square foot (the foot-candle) or lumens per square meter (the lux).

The intensity of light reflected from a surface is referred to as its surface brightness or 'luminance'. One unit of luminance is the foot-lambert, which is the luminance of a uniform diffuser emitting one lumen per square foot.

When a perfectly diffuse reflector is illuminated by one foot-candle and then has a luminance of one foot-lambert, its gain is said to be unity (one) in all directions. Unfortunately, there is no such thing as a perfect reflector, so normally the gain will be less than one.

However, sometimes a screen can have a gain of more than one. An individual point on a diffuse reflector radiates light in all directions; but if a surface reflects more light in one direction than another, the gain can be greater than one, *when measured in the specified direction*. For example, a screen made with a glass-beaded surface illuminated with 10 lumens per square foot might produce a luminance of 50 foot-lamberts when

viewed on the axis of projection, i.e. it has a gain of five. However, when viewed at an angle of 30 degrees off-axis, the illuminance is only five foot-lamberts, i.e. the gain is only 0.5.

High-gain screens can only be viewed from within a specified viewing angle. There are other restrictions on their use which will be described later.

Front-projection surfaces

There are various types of surface; some of the common ones are described here.

Matt white
This surface is the nearest approach to the perfectly diffuse reflector. The simplest matt white screen is a wall painted with matt white paint. More usually, however, a matt white plastic or PVC base is used. These screen surfaces have a gain of up to 0.9 on axis, with only a small fall-off at wide viewing angles (e.g. a gain of 0.75 at 35 degrees off axis).

Glass beaded
This was one of the first methods of achieving high-gain screens. Beads of glass are embedded in the surface; by internal reflection these ensure that most of the incident light is reflected back on or near the axis of projection. The tendency for the beads to discolor and even fall off, as well as difficulties with transportation, has meant that this type of screen is not now used.

Lenticular
The lenticular screen is a ridged silver surface which is able to produce similar results to the beaded screen, but without the problems. It is not quite so high-gain as the beaded screen, but is a very evenly illuminated screen within a specified viewing angle, beyond which the brightness drops off sharply.

Perlux

This is a trade name of Harkness Screens Ltd. Here, a matt white surface is sprayed with a special coating to give a medium-gain screen (the gain is about two on-axis) which can be viewed at all angles likely to be encountered in a real movie theater auditorium. It is the professional surface for movie theaters.

Daylight viewing

These are screens of extremely high gain but with a narrow viewing angle. They use embossed aluminium surfaces, both to give the high gain and, just as important, to reject light incident from other directions (e.g. light coming from the side or from above) from falling on the screen.

Choice of front-projection screen

Anyone planning a teaching area with only one projection source, and able to ensure a narrow viewing angle, can consider the use of screens for viewing in high ambient light. Most other users must ensure proper control of the lighting conditions.

Teaching and training areas with reasonable lighting control and single-source projection can use lenticular screens provided that the audience are within the optimum viewing angle. If this cannot be assured, then Perlux or matt white screens must be used.

Auditoria designed for projection (i.e. movie theaters, preview theaters, presentation rooms etc.) should use Perlux, except in those cases where matt white is indicated.

Multi-image projection presents a special problem. Normally matt white must be used to ensure even picture illumination. This is because, when a wide screen is used with many projection sources, any tendency to high gain on one part of the screen will necessarily mean low gain on another part. For example, someone on the left of the auditorium will see a very bright picture for the part of the image that is directly in front of him, but the image to his right will look progressively dimmer; what is worse, the screen will have a blotchy appearance. Matt white completely solves the problem. In permanent installations with reasonably long projection distances Perlux may also be used.

In permanent auditoria it is usually necessary for the accompanying sound to appear to be coming from the screen. A loudspeaker behind the screen would sound muffled because the screen surface will severely attenuate the high frequen-

cies. However, if the screen is *perforated* the sound is fine. In practice, a regular pattern of holes approximately 1 mm in diameter is used, with a total perforation of about five per cent of the screen area. This is not suitable for very small auditoria where the audience are close to the screen. Perforated screens can be obtained in both Perlux and matt white.

Special front-projection screens

Apart from video-projection screens, which are described separately below, there are three other special types of screen that may be of interest:

Deep curve

The curved screens used for some movie presentations can suffer from a lack of contrast due to light from one part of the screen scattering across to another part. Special coatings are available to minimize this problem.

3D

Projection of 3D movies is based on the projection of polarized images. Unfortunately, some conventional screens tend to destroy the polarization. Also, the polarizing process takes away a lot of the light, so a high-gain screen is needed. Special 3D screens, with multiple aluminum reflectors sprayed onto the surface, have an on-axis gain of about four. They cannot be viewed more than 25 degrees off-axis.

Very high gain

Front-projection screens with a viewing angle of only a few degrees are used in film and TV production. For example, if a film is made which

Figure 22.1 Motorized roll-down screens are suitable for training rooms.

involves action in some exotic location, it can save time and money to use a photograph as the background scene. One way is to use a projected backdrop while action takes place in front of it. No image is seen by the camera on any foreground objects. This is achieved if the projector and camera lens are both on exactly the same axis using half-silvered mirrors, which are actually specially coated glass.

Why back projection?

Although the optimum way of presenting a film or other projected AV program is by front projection in a darkened auditorium, there are many occasions when this is not possible. Very often back projection can solve the problem. Back projection is also the key to several applications of AV that are not auditorium based.

The majority of back-projection screens reject most light falling on the front of the screen. This means the image does not get washed out by ambient light. Back projection is, therefore, the key to those projection applications where the image must be viewed in high ambient light.

Back projection is essential in applications such as desktop slide/sound units and filmstrip units, microfilm viewers, etc. It also helps in small presentation rooms where AV material must be viewed with the lights on, to allow note-taking and discussion.

Another problem in some presentation rooms is the limited height of the room. Back projection can allow a greater image height than would be possible in a room where a front-projection beam might be obstructed by the audience.

Back projection is more flexible to design in to exhibition displays. It is also widely used in the theater and other live presentation applications where action takes place in front of a screen.

Back-projection screen material

There is a choice for different applications:

Glass
Ground or sandblasted glass of *no use* as a back-projection material. However, glass rear-projection screens where the diffusing surface is either applied as a coating or cast on are available. Glass screens are the most expensive, but also the most permanent. Big screens are difficult to transport. In presentation rooms they are best because they cut down sound from the projection equipment and are unaffected by differences of air

pressure on either side. A compromise is to use a plain piece of glass and a flexible plastic screen in front of it separated by less than an inch. CAT glass is a very fine-grain rear-projection glass screen, available in small sizes only, which is used for 'copy and title' work and even for video transfer, where the film or video camera is directed at the screen.

Acrylic
A wide range of acrylic plastic screens is available, from small fine-grain screens for microfilm viewers up to large screens for presentation rooms. The quality of the large screens varies from excellent to dreadful, so a demonstration is recommended before purchase. Most acrylic screens consist of a transparent acrylic plastic base with an applied coating. However, the best screens have both the diffusing layer and the pigment cast with the acrylic itself. This both gives better results and makes a more durable screen. A typical material of this kind is available in sheets up to 1500 mm × 2250 mm (5 ft × 7 ft), and has pigmentation to reject front incident light, a matt front surface to avoid specular reflections, and a diffusing layer to give a useful viewing angle of up to 40 degrees off-axis and a transmitted gain of 1.2.

The acrylic lens screens specially developed for video projection should definitely be considered for exhibition work where bright images are required and where their size limitations are acceptable.

Flexible translucent PVC
This is the most widely used material, especially for AV shows given to group audiences and in decor, etc. It is very easy to install; small pieces can be directly stapled into a wood frame. More usually the material is supplied with a strong plastic webbing around the edges, with grommet holes to allow lacing to a frame or with press studs for use with transportable screens. It is available in several different types, some suitable for folding without creasing, some not.

There is a trade-off between the amount of light transmitted and the degree of dispersion. For AV it is better to sacrifice some light on-axis in order to ensure wide viewing angle and minimum 'hot spot'.

The material is usually dark tinted to help with the rejection of surface incident light. However, white material is available, both for back projection in the dark and for the odd occasions when a surface is needed that will serve simultaneously as a back *and* a front projection screen.

It is possible to obtain flexible rear-projection material in seamless rolls up to 3 m (10 ft) wide.

Bigger screens can be made by welding the material. The weld seams are virtually invisible when a picture is viewed from a proper viewing distance.

A widely available general-purpose flexible rear-projection material has the following outline specifications:

☐ Vinyl latex base approximately 0.3 mm thick, and suitable for folding. It should be elastic, fungus resistant and fire retardant.

☐ Available off the roll in 3 m (10 ft) widths or made up to size, with or without edging, and with either grommets or press studs.

☐ Dark tint; transmitted gain on-axis approximately 2.5; low reflectance (about 15 per cent at 30 degrees); image resolution 20 line pairs per mm.

☐ Maximum recommended viewing angle 45 degrees off-axis; suitable for short focal length lenses.

A rear projection screen images the light from the projector. If wide-angle lenses are used not all of the light hits the screen at right angles. There is, therefore, a tendency to hot spot on back projection so that the center of the image appears significantly brighter than the edges. The problem is greatly reduced if *long* focal length lenses are used, but in back projection most users want to use short projection distances. In that case it is essential to use a material which is highly dispersive to minimize or eliminate the problem. In slide-projection terms, the highly dispersive material is suitable for 35 mm lenses, but 60 mm lenses should be used if space is available.

Screen construction

Screens are available in a wide range of executions to meet the needs of different uses and budgets.

Permanent installations should have fixed screens. Where one screen is to serve several different projection needs or formats, it should be fitted with motorized masking. Usually this is only side masking, but in some cases it is also necessary to adjust the top and bottom masking. Masking is done with a non-reflective fabric, usually a black wool serge. Many permanent installations also need a set of motorized curtains or drapes to cover the screen when it is not in use.

Lecture theaters and teaching environments may use fixed screens, sometimes covered by sliding panels, which may themselves be part of the teaching wall. Very often, however, the multi-purpose nature of the room demands the use of

Figure 22.2 Fold-up screens are best for travelling AV presentations. They pack into small boxes that easily fit in the back of a car.

ceiling or wall-mounted roller screens. Ideally, these will be motorized, but hand-operated screens are also available.

Travelling AV presentations need some form of portable screen. Within one building the most economical approach is the use of tripod screens for screen sizes up to about 2.5 m (8 ft) wide and box screens up to 3.6 m (12 ft) wide. Neither is particularly elegant, however, and neither is suitable for transporting by car. For these reasons most professionally made and presented AV shows that must be shown on the road use fold-up screens with a folding aluminum frame and screen surfaces that snap on with press studs. These screens are available in a wide range of sizes from about 1.8 m × 1.2 m (6 ft × 4 ft) up to 9 m × 3 m (30 ft × 10 ft) or even bigger. They are available with both front and back projection surfaces, and can also be supplied with dress-up kits to give a draped screen surround. The smaller sizes use a single-tube frame construction; larger ones, especially for wide-screen back projection, use a truss construction.

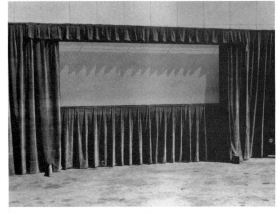

Figure 22.4 Temporarily installed screens can look better if they are fitted with a dress-up kit. (Photo The Screen Works Inc.)

Screens for video projection

Video projection presents some special problems. The Eidophor system of video projection is similar to a professional movie projector in its requirements, so a standard Perlux screen can be used. All the affordable projectors, however, give very little light, so every effort is made to extract the last lumen.

Some video projector units come complete with their own built-in screen system, either back or front projection. These will not be described because it must be assumed that the manufacturers have done their best to optimize the screen arrangement. These systems are characterized by a fixed lens-to-screen distance and screens that have as high a gain as possible. The front-projection screens are usually curved, and have a restricted viewing angle.

There are now several types of free-standing video projectors. All use three tubes (red, green and blue) and the total white light output is between 100 and 400 lumens, i.e. at best the output is less than half that of a standard slide projector.

Where it is possible to install a fixed screen that will *only* be used for the video projector, and where a relatively narrow viewing angle is permissible, it is preferable to install a high-gain screen. The best are rigid screen units with a curved aluminized surface. They are available in sizes up to 3 m (10 ft) wide.

Often this is not possible, perhaps because a wide viewing angle is required, or the screen system must also serve the needs of other AV shows, or the rigid screen is impractical. Then there is no choice but to use a standard screen. Both Stewart Screen of the USA (Ultramatte 300) and Harkness of the UK (Video Perlux) have special grades of material that may be useful.

Figure 22.5 The giant IMAX screen at the National Museum of Photography, Bradford, England. At 18 m (60 ft) wide this is a relatively small IMAX installation! (Photo Harkness Screens Ltd)

can be used, and will certainly give reasonable results in the dark. Special 'lens screens' are available that work on a fresnel lens principle, and have dark non-reflective striped front surfaces. These will give significantly better results when high ambient light levels are present; the difference is not so noticeable in the dark. Lens screens are very expensive, but may be essential in exhibition applications.

One type of video rear-projection screen is particularly worth knowing about, and is recommended both for exhibition work and small presentation rooms. It is now available in video format (4:3) but it is suitable for use with slide projection as well.

The Scanscreen™ video screen is supplied as a one-piece screen either 920 mm × 690 mm ('45-inch screen') or 1360 mm × 1020 mm ('67-inch screen'). The back of the screen is a large fresnel lens, which bends the incoming diverging light rays so that they all come through the screen at right angles to the plane of the screen. This solves the main problem of short focal length lenses whose projection beams diverge.

The front surface of the Scanscreen is another refraction system made up of parallel elements that are designed to achieve two things. The first is efficient dispersal in the horizontal plane to give good results up to 45 degrees off-axis. The second is to eliminate the color fringing that is often associated with video rear projection.

Because the video projector has three sources, a poor rear-projection screen will give three hot spots, red, blue and green. Even a good conventional material may show a tendency to color fringing, depending on the viewing angle. The special refractive elements used in the Scanscreen are specifically designed to eliminate this problem.

The result is a screen that gives an outstanding gain of 5.5 on-axis, with a gain of 2.7 even at 37 degrees off-axis (in the horizontal plane), making it suitable for use in well-lit presentation rooms. However, care should be taken to avoid lights pointing directly at the front of the screen.

Figure 22.6 The Barco Retrodata is a free-standing video-projection unit. It uses the Scanscreen 67 high-gain video rear-projection screen to ensure good results in lighted rooms. (Photo Cameron Communications Ltd)

Video projectors can also be used on back projection, although in principle only those giving 350 lumens or more should be considered for this purpose, except where a self-contained set is being used. Standard back-projection screen material

Chapter 23

The audio in audio-visual

High-fidelity sound in the home has made audiences aware of good sound, and they expect similar quality when they hear an AV show. Unfortunately the conditions under which AV shows are presented are not the same as those that apply in the living room, so the results are not always as good as expected. This chapter reviews a few topics that can help get the best out of AV sound.

Absorption

The average living room is quite small. Even a company presentation room is significantly bigger, and a hotel banqueting suite much bigger still. The result of this is that a sound system that sounds fine at home sounds very disappointing when played to an audience of any size. The sound seems to have lost its bite and words become difficult to understand. This is because the room furnishings, and in particular the audience, absorb the high and mid frequencies which give the intelligibility.

The first priority, as far as the audio in audio-visual is concerned, is that the message is heard. This usually means the use of loudspeakers that are larger than may be convenient. If it is not possible to take large loudspeakers to the presentation, then ones that are good on speech should be used.

There are special loudspeakers available that have been designed for auditorium and performance applications, and these should be used for all presentation work. These are of high efficiency and can project the sound to the back row.

Loudspeakers must be installed up high, for example on either side of the screen. If people can see the loudspeakers, they can probably hear them. If the speakers are on the floor the front row may be deafened but the back row may not hear a

Figure 23.1 Travelling AV shows should use performance loudspeakers at screen height. (Photo Tannoy Ltd)

thing. Many performance loudspeakers are available with lightweight telescopic stands that help position the loudspeakers correctly.

Speech reinforcement

It is surprising how even a small auditorium or meeting room needs a speech reinforcement or public address system. This is because of sound absorption by furnishings and people, and is very

Figure 23.2 The lectern with a built-in amplifier and speaker system is helpful in the small temporary venue. (Photo Edric Audio Visual)

noticeable in modern office-type environments with low absorbent ceilings and soft furnishings. Listener fatigue is reduced by always using a simple sound system.

For meetings that take place in rooms on an irregular basis, the portable lectern with built-in amplification can be helpful. However, because the lectern does not permit the loudspeaker to be sited very high, and because the proximity of the loudspeaker to the microphone limits the available sound level, these devices are best for small audiences in temporary meeting rooms.

Permanent sound reinforcement systems in presentation or meeting rooms with low ceilings will often use ceiling loudspeakers. Although this has the disadvantage that the sound no longer seems to come from the presenter, it has the advantage of giving high speech intelligibility while allowing the system to be operated at an unobtrusive level.

Larger presentation rooms can be treated more like theaters, with loudspeaker systems designed to give even coverage to all the audience while

maintaining the illusions that the sound is coming from the stage. In deep auditoria or large presentation rooms it may be necessary to have additional loudspeakers placed further down the auditorium to get good coverage at the back without the sound being too loud at the front. This is especially true when trying to reach people in seats under a balcony. Until relatively recently the problem was that people sitting at the back not only had the illusion that the sound was coming from very near them, but also often heard the front loudspeakers as an 'echo' because the sound from the front loudspeakers was reaching them appreciably later than that from the nearby loudspeakers. This problem is now completely solved by the use of electronic time-delay units. These delay the sound signal which is fed to the back loudspeakers by the necessary few milliseconds to ensure that the sound from the distant loudspeakers arrives at the listeners' ears at exactly the same time as that from the near loudspeakers. This not only eliminates the 'echo' problem, but also helps maintain the illusion that the sound is coming only from the stage.

Another special sound reinforcement problem is common in round-table conferences and large boardrooms. Often the number of participants is so great that a person at one end of the table can have real difficulty hearing what is being said by someone at the other end. The problem can be solved in two ways. The lowest-cost method, and one which is quite suitable for temporary installations, is to equip each delegate with a combined microphone/loudspeaker unit. These units are arranged so that if a particular microphone is switched on, its associated loudspeaker is switched off to prevent acoustic feedback. Otherwise all loudspeakers receive the same signal from whichever microphone is in use.

Switching is sometimes manual, arranged so that the action of switching on any one microphone automatically switches off the one previously in use. Generally, the chairman has an arrangement whereby he cannot be overridden. More recently automatic switching, where the presence of a voice input signal automatically switches on the microphone, has been introduced. This arrangement is a standard feature of the second, more expensive, system in which the automatic sound reinforcement system is built in to the boardroom table. The design goal is a system which is totally unobtrusive in use, requires no operating by either an operator or the participants, and ensures that everything that is said by any participant is clearly heard by all the others.

Figure 23.3 Conference microphone systems use special switching techniques to ensure that every delegate is clearly heard. (Photo Electrosonic Systems BV, courtesy NLB Bank, Amsterdam)

Noise reduction

When a tape is played to a large audience, they can be much more aware of its imperfections than they would be when listening to it at low volume on their own. Most professionally produced AV shows have original sound tracks of high quality, so it seems a pity to lose this quality at the final replay. On tape-based systems noise reduction should always be used to eliminate tape hiss and noise. The best known is the Dolby™ system; others are dBx™ and High Com™. Some other points to remember are:

☐ Only tape and tape cassettes of the highest quality should be used. It is a false economy to use substandard cassettes.

☐ Unsuitable tapes must be avoided. In theory metal and chromium dioxide tapes can give better results, but only if the replay equipment is suitable.

☐ Wherever possible 'standard play' tapes should be used. Sometimes the use of 'long play' tapes is unavoidable, but 'double play' tapes should *never* be used for AV applications.

☐ A spare copy of the show tape must always be available.

☐ Show copies must *never* be made on non-professional tape duplicators. These machines are suitable for mass-copying tapes that will be listened to by an individual, but not for an important show that is to be heard by an audience. Some duplicators have such a poor frequency response that they cannot properly copy control signals which are only at 5 kHz.

Multi-track tapes

Straightforward commercial AV shows need no more than a good quality mono sound track. Slide and multi-image shows presented to large audiences, especially those designed for motivation, promotion or entertainment, benefit from stereo sound. There are some shows that use more than two sound tracks.

For convenience many small shows are run from standard compact cassettes. The track format of these cassettes was originally designed to meet the needs of the domestic user, and they have two pairs of tracks each intended for a stereo audio program. For this use it was important that there was negligible crosstalk between the two programs, but it did not really matter if there was crosstalk within the stereo pair.

As soon as cassettes started being used for AV it was found to be impractical to use one half of the stereo pair for audio, and the other for control signals. When the level of the control signal is loud enough to reliably operate the programming

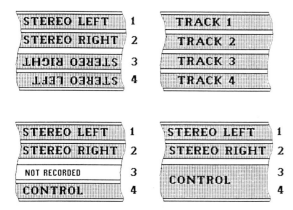

Figure 23.4 Different track standards on the compact audio cassette.

device there will be objectionable interference on the audio track. Therefore, cassettes used for AV are only used in one direction and the control signal is placed where the second stereo pair would normally be.

Unfortunately, there has been some confusion as to the precise placing of the control track information. As far as international standards are concerned, tape cassettes run at 4.75 cm/s, and the tracks are placed as shown in Figure 23.4. However, to meet the needs of the musician market some manufacturers have introduced cassette machines with four equally spaced tracks and a speed option of 9.5 cm/s. There is therefore a tendency to produce multi-image show copies with the control signal in the position of track 4 on the equi-spaced system, rather than in the standard mono position covering both tracks 3 and 4 of the two-pair system. Care must therefore be taken to ensure that show-copy tapes are prepared in a manner that is suitable for the intended replay equipment, because control signals recorded on the 'track 4' basis may be marginal when played back on equipment intended for 'AV standard' replay.

Tape cassettes are normally only used for show replay or in the production of very simple slide/sound sequences where copies are not required. Reel-to-reel tape is used for all main show production work, and for prestige show playback systems, both to ensure better quality and to allow editing of the tape. In fact, the frequency response available from modern tape cassette machines matches that of reel-to-reel tapes, but reel-to-reel gives better overall results because the larger tape area and higher tape speed is less susceptible to drop-outs. Physical editing of

the tape is still done on master mono tapes, but this has largely given way to the use of cross-dubbing in the assembly of a final show. Cross-dubbing has become feasible because of the introduction of multi-track tape machines which allow the different elements of a program to be laid down side by side for subsequent mixing down. Modern professional recording studios use 16 and 24 track machines.

Figure 23.5 A typical multi-track tape recorder used in professional sound recording studios. (Photo Soundcraft Ltd)

Developments in the sound recording business have led to the introduction of high-performance and reasonably priced multi-track tape recorders. Those of most interest to AV users are:

☐ 4-track quarter-inch, the workhorse of the AV business.

☐ 8-track half-inch, the standard for the big show.

☐ 8-track quarter-inch, a new format with interesting possibilities.

Multi-image shows require at least one control track to regulate the activities of between 15 and 56 projectors (see Chapter 16). Depending on the method of production they may or may not need a clock track as well as, or instead of, the control track.

Wide-screen shows benefit from three tracks of audio to ensure a good center sound channel; conventional stereo can result in a 'hole in the middle' sound. Putting the main commentary on a separate center channel enables the sound to be rebalanced to ensure that the commentary is heard properly if there are any acoustic difficulties.

Figure 23.6 Sound system for a medium-size public entertainment show using four-track tape. (Photo Tony Gidley Productions)

Big shows also use effects tracks played through loudspeakers at the back of the auditorium, and perhaps from other sources to locate the sound. Sometimes programmed sound-switching is used as well as the multi-track technique. A feature of some entertainment shows is the introduction of enhanced bass to improve the effect of full-range orchestral music and sound effects such as trains, explosions etc. This bass can either be derived from all the sound tracks via a special filter, or can be carried on a track of its own. The loudspeakers used are very large and not really suitable for the travelling show! Most loudspeaker systems can give a reasonable performance down to about 50 Hz, but to achieve the visceral impact of these special effects requires the use of loudspeakers able to deliver a strong acoustic output at frequencies down to 18 Hz. To do this, drive units of up to 75 cm (30 in) diameter are used.

Another obvious use of multi-track sound is to carry different languages. These can be played through the main system as alternatives to the first language, or can be fed via headsets to those members of the audience who need the alternative. This means that either the seats must be wired with headphone sockets and appropriate controls, or visitors must be given suitable wireless receivers with transmission by induction loop or infra-red.

Figure 23.7 The sound replay system at the Royalty and Empire exhibition at Windsor, England. Note the use of two eight-track tape decks.

Figure 23.8 At The London Experience alternative-language commentaries can be heard via individual loudspeakers.

If only a proportion of the audience needs to hear the other languages, the alternative is to fit the back row of seats with headrests with built-in low-level loudspeakers. The system should then be designed so that at the end of the show all of these loudspeakers that have been in use are automatically switched off.

Tape recorders used in permanent installations can be fitted with automatic rewind control to allow for fully automatic performances. Prestige and revenue-earning shows are always fitted with two tape decks with a simple changeover device, both to ensure proper back-up in the event of tape deck failure, and to allow tape deck maintenance to be carried out while the show is in progress.

Super sound systems

Audio is still to some extent a black art beset with a lot of jargon and riddled with personal opinion. What *is* clear is that the price of sound equipment goes up exponentially: to get a significant improvement over a $10,000 system, you might have to spend $40,000.

When designing a sound system for AV presentations a responsible sound contractor will start by defining the basic requirements:

☐ How many sound tracks?

☐ What are the other sound sources?

☐ What is the size and shape of auditorium?

☐ How big is the audience?

☐ How often does the show run?

☐ Is automatic or manual control needed?

☐ What is the required sound pressure level?

There is now a good choice of equipment available to fulfil the requirements of each part of the system, and it is relatively easy to produce a specification that will give good results at a reasonable cost in most circumstances. It is not uncommon to encounter specifications calling for sound pressure levels so high that if the system were actually run at the full level the audience would suffer permanent damage to their hearing. The problem about such specifications is that the user ends up spending unnecessary money on equipment that might have gone towards improving the show!

The sound level needed will depend on the expected program content and the nature of the audience. There must then be some headroom to allow the system to deal with transient peak levels without distortion. A technique used to improve the smoothness of sound is graphic equalization, where room resonances and unexpected absorption peaks, as well as the non-linear response of the loudspeakers, can be compensated for electronically. This is well worth having in difficult auditoria with hard reflecting surfaces, but the system must be set up properly. There are probably more shows whose sound has been spoilt by badly set-up correction equipment, than ones which have benefited from it.

The aim of graphic equalization should be to end up with a system that gives a totally predictable performance. Any program material fed into it should be reproduced without any peaks or troughs in the frequency response. This is achieved by feeding into the complete sound chain a 'pink noise' signal which contains all the frequencies required to be reproduced at the same level. A microphone at the intended listening position picks up the signal that comes out of the loudspeaker and feeds it to a spectrum analyser. This device displays how much of each frequency is being received. In the ideal system the display would be a straight line; in practice, it will look like a small mountain range. The graphic equalizer is the device in the sound chain that is able to selectively boost or attenuate narrow bands in the

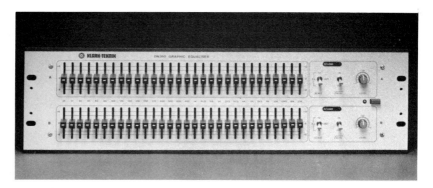

Figure 23.9 A graphic equalizer can compensate for difficult acoustics in an auditorium. (Photo Klark Teknik Ltd)

audio spectrum. Typical professional equalizers allow adjustment of 31 different frequency bands with their center points at one-third octave spacing. The equalizer is adjusted until the spectrum analyser shows as near a linear response as possible.

If any end user is in doubt where to spend his money, it should be spent on good loudspeakers first, then on tape decks and amplifiers, and lastly on fancy correction equipment.

Digital sound

Most audio tapes are analog, that is, they record an electrical analog of the original sound. In digital sound the electrical signal is instantaneously measured and turned into a digital signal with values of 0 or 1, and coded to represent the instantaneous level. All subsequent processing is done digitally until the final replay when the digital signal is decoded. The method is technically very complex, but made possible by recent advances in electronics. Its potential advantage is that noise of all kinds is virtually eliminated.

The quality of the digitally encoded sound depends on two factors. The frequency response depends on the number of samples taken per second. Mathematically, it can be shown that the sample rate needs to be double that of the maximum frequency required. Thus, if a bandwidth of 10 kHz is required, it is necessary to take 20,000 samples per second. The dynamic range of the sound depends on the number of bits per sample. If there are only eight possible values of sound level, which is all that can be achieved with three-bit sampling, the resulting sound will either be monotonous or will jump in level. A minimum practical number is eight-bit sampling (256 sound levels).

Digital sound in the form of solid-state sound stores will have an increasing impact on permanent sound replay systems (see Chapter 10). For speech and simple sound effects a system using eight-bit samples (aided by a digital compression and expansion technique that gives improved dynamic range at low sound levels), with a sampling rate of 8000 per second, is sufficient. This gives a bandwidth of about 4 kHz, a performance similar to AM radio without any extraneous noises, and is reasonably economic on memory, needing 64,000 bits to be stored per second. The present generation of 256K memory chips can thus store four seconds' worth of sound. Applications which need better frequency response must use more memory and can typically give either a 7.5 kHz or 11 kHz response.

For high-fidelity applications it is necessary to use an order of magnitude more memory. These typically use a sample rate of 42,000 per second, and thirteen-bit encoding. Master recordings are made on magnetic tape, which requires a new generation of multi-track digital recorders. Playback systems can no longer use chips to store the sound, because this would be ridiculously expensive and inconvenient. The compact audio disk is the best known storage medium and it is likely to take over the position held by the LP record. It uses the same technology to store digital audio data as the laser videodisk.

However, the same digitial technique can also be applied to magnetic tape. Until recently this was only applied to the special tape recorders used for making professional master tapes. Now it is set to form the basis of a new consumer medium. Digital Audio Tape (DAT) uses a tape cassette of

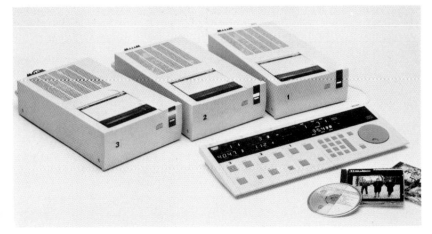

Figure 23.10 The Philips professional compact audio disk player. Usually used in broadcasting and professional audio studios, it could also be used in special show playback systems.

Figure 23.11 The Sony PCM 701ES allows the recording of a stereo digital audio sound track on a standard videocassette recorder. (Photo HHB Hire and Sales Ltd)

small dimensions to give the same initial quality as CD. The system has been defined, but has not yet (at the time of writing) been released on the market.

Although the compact disk is highly acceptable as a consumer product, it is not very practical for the AV user. The compact disk has a time code facility, so it is theoretically possible to run a digital audio program in synchronization with an AV show, but the cost and inconvenience of having a single disk made rules it out for most users.

There are two ways in which high-fidelity digital audio can be applied to AV. The first is suitable for roadshow work which requires high-quality stereo sound. This can be achieved by encoding the digital data directly on to a standard videocassette recorder, i.e. the wide bandwidth of a video recorder is used, not for a television picture, but for storing the enormous amount of digital data needed to represent high-quality audio. The equipment needed to do the encoding and decoding is relatively inexpensive, and what would have been the audio tracks on the videocassette can now be used for AV control.

Thus, in theory the show producer who uses a digital audio studio to produce his master tape can now take digital sound on the road.

When DAT is established, there may be machines suitable for AV work. There will need to be some method of extracting a time code reference for synchronization.

Another likely development will be of more use to the permanent show installation. In the same way that solid-state sound stores will replace small tape-replay devices for continuous duty in simple commentary and effect applications, the recordable optical disk may well take over the role of multi-track analog tape recorders for 'big show' replay. The optical data disk will soon have a major impact in computer applications. It uses similar technology to the laser videodisk and it can record huge quantities of data. There is no reason why this should not be digitally encoded audio. One 30 cm (12 in) diameter disk could comfortably hold all the data for a big show, for example a show 40 minutes long with four high-fidelity sound tracks, four alternate language tracks to be heard via headphones, and all the computer control data needed to run the show.

Glossary

The brief glossary that follows includes most of the technical terms likely to be encountered on first acquaintance with industrial AV. Many are explained in much more detail in the relevant chapters. The definitions and descriptions are those appropriate to industrial AV. For those who require a more comprehensive glossary the BKSTS *Dictionary of Audio-Visual Terms*, published by Focal Press, is recommended.

Aerial image An optical image formed in space rather than on a screen. It is used in creating film special effects and in slide or film to video transfer. In film to video transfer a group of projectors work through an *optical multiplexer* to form a combined *aerial image* which is received by the video camera.

anamorphic An anamorphic lens squeezes a picture to obtain a wider aspect ratio. See Chapter 21.

answer print The first print of a complete motion picture, including the sound track. It is normally used as the basis for making decisions about detailed color correction on the subsequent prints.

aperture In projection terms the aperture is the size of the original image being projected, e.g. 35 × 23 mm for 35 mm slides.

ASCII An acronym for American Standard Code for Information Interchange. There is a simple digital code for all upper and lower case letters, numbers, punctuation marks etc. This code is used by computers to communicate with printers and other peripherals.

aspect ratio The ratio of the width to the height of an image. Slides are usually 3:2. See Chapter 17 for different film aspect ratios.

back projection This is when the projector is placed behind the screen, and the audience views the image by *transmission* of light, as opposed to *reflection* when front projection is used. See Chapter 22.

BASIC An acronym for Beginner's All Purpose Symbolic Instruction Code. It is the best-known language for writing computer programs. It is mainly used by beginners, and is also used by experts when programs must be written quickly, but where speed of execution is not important.

baud The rate at which computer data is transmitted, usually in bits per second.

bit The fundamental unit of computer information. It can only take the values of 0 or 1; all calculations within a computer are carried out using this binary notation.

BVU A trade name of Sony; Broadcast Video U-matic, otherwise known as high band U-matic. It is a three-quarter-inch videotape system based on the use of U-matic videocassettes, but it has a greatly extended frequency response and an overall performance that is up to broadcast standard.

byte A group of bits. An eight-bit byte can represent all numbers up to 256.

Carousel A registered trade mark of Kodak. It describes their range of automatic slide projectors using a circular slide tray and gravity slide feed.

cartridge In AV it is likely to be encountered (1) as the pick-up or cartridge on a phonograph or record player; (2) as a device to hold an endless loop of quarter-inch tape in continuous sound playing devices; (3) as a device to hold both an endless audio tape and a film strip in desktop AV filmstrip units; (4) as a device to hold an endless loop of Super 8 movie film.

cassette The tape cassette was devised to make tape handling easier. The tape is enclosed with its supply and take-up reels. The standard audio cassette is the CC cassette, and is available with different lengths and qualities of tape. AV shows should use tape of the highest quality, and should use standard play when possible, *never* double play. See also *videocassette*.

chip The popular name for an integrated circuit, a single component which itself contains up to many thousands of transistors and other components which have been produced on one substrate by a miniature lithographic process. Components are produced by selective etching and diffusion with trace elements.

compact audio disk A new means of sound recording that uses videodisk technology to record digital audio signals. It may eventually supersede the conventional phonograph record. It is of little relevance to industrial AV, although in principle there is no difficulty in synchronizing audio disks with multi-image and other AV shows.

composite video An electrical signal carrying the information for an entire color television program, but not including the sound. See also *RGB* and *RF*.

computer graphics Computer graphics are now playing an increasing part in visual aids. Many companies offer a service where high-quality color slides are first made on computer and are then automatically transferred to film with high resolution. Lower-cost methods include direct display of computer graphics on monitors or by video and data projectors. It is even feasible to make instant slides directly from a computer color graphic terminal.

data projector The output of a computer that would normally be shown on a monitor can be fed to a video projector. However, the resolution of these projectors is not always good enough for video applications. Data projectors are video projectors of high bandwidth. Monochrome data projectors are really only suitable for computer data. The color projectors are also suitable for other color video programs. They use an RGB signal.

digital In electrical terms signals can be digital or analog. Most audio recordings are analog, i.e. the signal on the tape varies in the same way as the sound to be reproduced. In a digital signal there are only two values, on or off, 1 or 0. Digital audio recordings take the instantaneous level and code it as a binary number. This type of system is greedy in the amount of data space it needs, but theoretically, it gives almost noise-free recording with no degradation on copying. Digital message repeating systems are likely to replace tape in heavy-duty applications. All computer signals are digital, so the word often appears in references to computer and programming equipment. Digital processing is also applied to video signals.

dissolve unit Controls two automatic slide projectors to dissolve one image to another using electronic lamp control. See Chapter 13.

Eidophor Trade name of Gretag AG to describe their powerful video projector which uses a light-valve technique. See Chapter 19.

episcope Otherwise known as an opaque projector. It is a device for projecting directly from, for example, the pages of a book. It is not used much now because it is easier to make an overhead transparency or slide instead. An **epidiascope** is a device that combines the facilities of an episcope with those of a slide projector.

equalizer A device for deliberately modifying the frequency response of an audio system, in other words a sophisticated tone control. *Graphic equalizers* allow individual frequency bands to be boosted or attenuated; there are typically as many as 32 bands in a one-third octave graphic equalizer. This device is used to match an audio system to an auditorium by reducing frequencies that have a resonance and boosting frequencies with exceptional absorption.

feedback The feeding of an output signal back into the input. Some kinds of feedback are beneficial, but the one most AV users are likely to encounter definitely is not! Acoustic feedback is when a microphone picks up its own amplified signal from the loudspeaker, causing a howl. The cure is to turn the volume down, but also to ensure that the loudspeakers are forward of the microphone and pointing away from it.

filmstrip Individual images intended for still projection printed on a continuous length of film, usually 35 mm, but also 16 mm. It is cheaper than slides, especially when a lot of copies are needed, but not as flexible. It is used in industrial AV for individual communication devices which often use filmstrip cartridges for ease of operation.

flipchart A very large pad of paper, usually supported on an easel. The presenter uses it as a disposable writing board, or prepares a number of pages with information and flips them over as they are described. Writing is by colored marker pens. Flipcharts are useful for meetings and informal presentation, especially where facilities are limited.

floppy disk A flexible magnetic disk for storing computer data and computer programs. It was originally 8 inches in diameter, but now most microcomputers use 5¼ inch disks, some a protected 3½ inch disk.

focal length A length, usually expressed in

millimeters, that relates the image size produced by a lens to the distance of projection. See Chapter 21.

frequency Cycles per second or hertz (Hz) as applied to electrical and audio signals. Most audio information is in the range 20–15,000 Hz. Note also that 1000 Hz equals 1 kilohertz (kHz) and 1,000,000 Hz equals 1 megahertz (MHz). A TV signal on videotape needs about 3.5 MHz. VHF broadcasts are at around 100 MHz.

genlock All video devices require a source of *sync pulses* that determine the precise moment at which a video frame starts. When several devices, e.g. a video recorder and several cameras, are being used together, it is vital that their sync pulse generators work in unison. This is achieved by genlock.

hard disk Although floppy disks are exchangeable, most hard disks are not. However, they hold much more data, and can be accessed much quicker. Winchester disks are hard disks.

HDTV High-definition television. At present much research is going on to introduce a television system of higher definition. The problem is to design a system that does not use too wide a frequency response, and that is preferably in some way compatible with existing systems. In Japan there is already an experimental 1125-line system with a picture ratio of 5:3.

holography A means of describing an image in terms of the way it interferes with light, an abstruse concept that is not easily compressed into a few words. The *hologram* is the holographic equivalent of a photograph. Holograms are interesting because they store *three-dimensional* images, and because if a hologram is broken each part of it can still reconstruct the whole image, albeit at a lower definition. Holograms require laser light for their creation and usually for their viewing. They are suitable special exhibition displays, but not, as yet, for projection. Holography is unlikely to be relevant to AV for some time yet. There are some cases where an audience believe they have seen projected holograms, when actually they have seen conventional film or video display viewed indirectly via partially reflecting glass. This is the 'Pepper's ghost' technique which is very effectively used by, for example, displays in Walt Disney World.

instant slide Presentations have a habit of needing visuals prepared at the last minute. The overhead transparency can be made very quickly, but until recently 35 mm slides needed a significant processing time and special facilities.

Now Polaroid has an instant slide film requiring no special facilities, and with a development time of a few minutes.

interface A computer cannot be directly connected to the outside world. For example, if a computer had to switch on a light bulb it has no circuit element strong enough for the task. Even if it could, the connection would introduce unwanted interference into the computer. To protect it from the outside world a computer uses interfaces to isolate both inputs and outputs. Interfaces allow computers to become an active part of a control system, and are often themselves highly sophisticated devices.

keystone When a projector of any kind is not able to project on-axis the image projected is not rectilinear and suffers from keystone distortion. With overhead projectors the problem is eliminated by tilting the screen. With multiple-slide projectors the problem is reduced by keeping the projectors as close together as possible, or by the use of special lenses.

laser An acronym for Light Amplification by Stimulated Emission of Radiation. A device that produces an intense narrow beam of coherent light (i.e. the light waves are all in step), usually of only one or two wavelengths (colors). It is of very little use in AV except as a gimmick at product launches and similar events. In those cases the system must be under the control of an experienced operator. Baby laser displays are disappointing. However, one useful low-power device is the laser pointer, a battery-operated device ideal for lecturers in big halls. Lasers are also used in creating and viewing holograms. Compact lasers form the basis of precision scanning devices, for example in videodisk players. Experiments are proceeding in using lasers as a means of directly projecting large video images.

lumen The measure of light emitted by a light source, e.g. a 250 W projector lamp emits about 6000 lumens in all, but in a slide projector only about 1000 lumens will reach the screen because the projector optics are only able to collect a proportion of the light. The **lux** is a unit of illumination, lumens per square meter.

machine code High-speed computer programs are written in machine code, the actual internal language of the computer. Each instruction accomplishes a tiny task and relates to the manipulation of individual bits and bytes.

magnetic sound Here sound is recorded in the form of magnetic patterns in a magnetic powder applied to some base material. Most AV sound is magnetic, whether it is on cassette or open-reel tape. Magnetic sound can also be applied to movies, in which case one or more magnetic stripes are bonded to the edges of the movie film.

memory The power of a computer is measured not only by its speed but also by the size of its memory. This is divided into high-speed working memory or RAM (Random Access Memory), typically 64–256 kilobytes, and the back-up store which is on floppy or hard disk. This can be anything from 100 kilobytes to many megabytes. A kilobyte equals 1024 bytes.

microcomputer The first computers were big mainframe machines; then came minicomputers, widely used in industry, most recently microcomputers have exploited the power of the microprocessor. The problem is that present-day micros are more powerful than yesterday's minis or even than the day before yesterday's mainframes. Let's just call them computers!

microprocessor An integrated circuit or chip that embodies all the main attributes of a computer. In fact, there must be several additional components around the microprocessor to interface with the outside world. There are many different kinds of microprocessors, some powerful sixteen-bit ones suitable for business computers, others with lots of interface capability and their own internal memory suitable for industrial control applications.

monitor A precision TV set for monitoring picture quality. The word now tends to be applied to any TV set with a video, as opposed to *RF*, input. A receiver monitor is a set that is both a conventional TV set for receiving broadcast TV, and has a video input.

multi-image The term preferred in North America for **multivision.**

multimedia Shows using more than one medium, especially mixed movie and multivision, perhaps with synchronized lighting control. Shows using mixed video and multivision are also possible.

multivision The programmed activities of many slide projectors on one or several screens.

noise reduction Most sound recording systems introduce noise (hiss) to the signal. This can be subjectively reduced by using a noise reduction system that effectively boosts the quiet part of a program for recording purposes, but restores the signal to the correct relative level on replay. The best-known system is the Dolby™ system; others are dBx™ and High Com™. Professional AV shows should use noise reduction.

NTSC See *PAL.*

optical sound Most 16 mm and 35 mm movies have optical sound tracks, where the sound is recorded photographically and replayed by a photocell. Modern 35 mm sound tracks are good, especially when used with a noise reduction system, but 16 mm sound tracks are less so, because they have a restricted frequency response. However, for most applications they give acceptable results and they represent the most reliable method of sound synchronization for movie.

overhead projector A device for projecting large transparencies of page size. It consists of a lamp surmounted by a fresnel lens stage on which the transparency, or transparent foil, is placed. A lens with an angled mirror above the stage focuses the image on a screen which is usually above and behind the presenter. The size of the original image allows very bright images on the screen.

PAL Unfortunately TV signals are not standardized throughout the world, and there are even variations within each main standard. PAL is the standard in Europe, except in France where they have their own SECAM system. NTSC is used in North America and Japan.

presentation unit Combines the facilities of a dissolve unit with those of an audio-visual tape recorder.

programmer Device to program the activities of one or more slide projectors, usually in synchronization with a sound track. See Chapter 16.

PROM An acronym for Programmable Read-Only Memory. When a microprocessor is used as part of a control product, e.g. a dissolve unit, its program is stored in a solid-state read-only memory. These can be specially made for the job in hand, but for many applications it is easier to buy a blank memory from the chip maker, and program it as required using a PROM programmer. There are various kinds of PROM. Possibly the most useful is the EPROM, which can be erased by intense ultra-violet light but is otherwise non-volatile, i.e. it does not lose its memory when the power fails.

pulse The control signal on magnetic tape used to control slide projectors. In simple slide/sound systems a 1 kHz pulse of about half a second duration is recorded to advance the slide. Multivision systems use a continuous stream of digital pulses of great complexity.

random access The ability to locate rapidly and show a visual aid or AV program. See Chapters 12 and 13.

reel-to-reel Conventional magnetic audio tape is supplied on reels, the tape being threaded through the record/play mechanism to an empty take-up reel. All master audio tapes are made reel-to-reel because it is easiest for editing and gives the best sound quality.

relative aperture   The ratio of a lens diameter to its focal length; also known as the *f*-number. See Chapter 21.

RF   Broadcast television uses signals at radio frequency or RF. This is the kind of signal picked up by a TV aerial (antenna). It usually includes the audio information as well. It is created by using a composite video signal to modulate a very high frequency signal suitable for transmission.

RGB   Color television signals are originated as three separate pictures: red, green and blue. Usually they are merged together as a composite video signal, but for the highest quality and for specialist applications like data projection, the signals are kept separate. Not all monitors can accept the separate RGB signals.

rostrum camera   Copying camera used in making professional AV shows. See Chapters 2 and 4.

SECAM   See *PAL*.

slide/sound   The French use the word diaporama, which sounds much more grand. Here it is taken to mean single-screen AV shows based on slides with a synchronized sound track. It can grow into multivision.

soft-edge mask   Used in multivision to allow projection of panorama pictures from several projectors, without any obvious picture join showing. See Chapter 14.

stereo   The recording of the sound signal onto two tracks of tape to give a sense of space and direction to the sound. Prestige AV shows use stereo and some use three or four tracks for special effects or to give a good center sound.

sync pulse   See *genlock*.

teaching wall   The well designed training room uses a teaching wall that neatly integrates different communication methods – sometimes side by side, sometimes using a system of sliding panels. Thus, in one permanent assembly the facilities of flipchart, white board, slide/movie screen and overhead projection screen can be combined.

teletext   The generic term for a system that presents alpha-numeric and graphic data as still video images. See Chapter 19.

three-dimensional projection   At present the only practical way to provide projected three-dimensional (3D) images is by the long-established method of taking two photos of the subject, with the two camera lenses separated by approximately the spacing of the human eyes. The images are then projected by two projectors onto the same screen. Each projector has a different polarizing filter. The viewer wears a pair of special glasses with a different polarizing filter for each eye. Each eye then sees only the image intended for it. The principle works equally well for movie film and for slides, and if care is taken with production the results can be spectacular. This technique is of rather limited value in the AV field, because few industrial and commercial subjects can benefit from 3D, other than as an occasional gimmick. Experimental systems have been demonstrated that use special lens screens to allow 3D images to be seen without glasses.

time base correction   A method of ensuring that a number of separate video devices all work correctly in picture sync. See Chapter 19.

track   Referred to magnetic tape. A tape can have more than one independent track on it, thus giving the possibility of stereo sound, or sound plus control. The normal audio cassette has two pairs of tracks. Typical reel-to-reel recorders have four tracks on quarter-inch tape or eight tracks on half-inch tape. Studio recorders have up to 24 tracks on two-inch tape. Typical videocassette recorders have two conventional audio tracks and the helically scanned video track.

transparency   Slides are transparencies, i.e. color photographs taken on 'reversal' films. In the AV field, however, the word normally refers to the transparency used on overhead projectors, which is nearly page size.

U-matic   A type of *videocassette* used for professional production and for industrial video shows.

video projector   Usually TV images are viewed directly on a cathode-ray tube, but there is a practical limit to how big these can be. Bigger images can only be produced by projection. See Chapter 19 for a description of the systems available.

video recorder   Device for recording TV programs. Video recorders used in broadcasting use 2-inch or 1-inch tape on reels. AV use is based on *videocassettes*. All video recorders at present use multiple tape heads to scan the tape diagonally in order to achieve the high tape-to-head speed necessary to record the high-frequency signals.

videotape   Magnetic tape designed for recording TV programs; recording is done by a videotape recorder.

videocassette   The handling of videotape is inconvenient, so it is normally presented in a cassette. The three-quarter-inch U-matic cassette is used for industrial and high-quality work. The half-inch cassettes used in the home, VHS, Beta and V2000, are increasingly also being used for industrial AV.

videodisk A means of storing video programs rather like a phonograph record. See Chapter 12 and 19.

writing board The old school blackboard has come a long way. Some applications are still best met by glass chalkboards, especially in teaching rooms. Other training rooms and some pre-sentation rooms use plastic or metal whiteboards with colored marker pens.

zoom As applied to lenses: variable focal length. In camera lenses this gives the ability to take close-up and distant shots with the same lens. In projection lenses it allows the projectors to be placed conveniently, with the lens adjusted so the picture exactly fills the screen. Some video camera lenses have zoom ranges of in excess of 10:1. Projection lenses usually only have a range of about 1.8:1.

Index